CW01151509

▶ **Marketing Big Oil**

Other Palgrave Pivot titles

Nicholas Robinette: **Realism, Form and the Postcolonial Novel**

Andreosso-O'Callaghan, Bernadette, Jacques Jaussaud, and Maria Bruna Zolin (editors): **Economic Integration in Asia: Towards the Delineation of a Sustainable Path**

Umut Özkırımlı: **The Making of a Protest Movement in Turkey: #occupygezi**

Ilan Bijaoui: **The Economic Reconciliation Process: Middle Eastern Populations in Conflict**

Leandro Rodriguez Medina: **The Circulation of European Knowledge: Niklas Luhmann in the Hispanic Americas**

Terje Rasmussen: **Personal Media and Everyday Life: A Networked Lifeworld**

Nikolay Anguelov: **Policy and Political Theory in Trade Practices: Multinational Corporations and Global Governments**

Sirpa Salenius: **Rose Elizabeth Cleveland: First Lady and Literary Scholar**

Sten Vikner and Eva Engels: **Scandinavian Object Shift and Optimality Theory**

Chris Rumford: **Cosmopolitan Borders**

Majid Yar: **The Cultural Imaginary of the Internet: Virtual Utopias and Dystopias**

Vanita Sundaram: **Preventing Youth Violence: Rethinking the Role of Gender and Schools**

Giampaolo Viglia: **Pricing, Online Marketing Behavior, and Analytics**

Nicos Christodoulakis: **Germany's War Debt to Greece: A Burden Unsettled**

Volker H. Schmidt: **Global Modernity. A Conceptual Sketch**

Mayesha Alam: **Women and Transitional Justice: Progress and Persistent Challenges in Retributive and Restorative Processes**

Rosemary Gaby: **Open-Air Shakespeare: Under Australian Skies**

Todd J. Coulter: **Transcultural Aesthetics in the Plays of Gao Xingjian**

Joanne Garde-Hansen and Hannah Grist: **Remembering Dennis Potter through Fans, Extras and Archives**

Ellis Cashmore and Jamie Cleland: **Football's Dark Side: Corruption, Homophobia, Violence and Racism in the Beautiful Game**

Ornette D. Clennon: **Alternative Education and Community Engagement: Making Education a Priority**

Scott L. Crabill and Dan Butin (editors): **Community Engagement 2.0? Dialogues on the Future of the Civic in the Disrupted University**

Martin Tunley: **Mandating the Measurement of Fraud: Legislating against Loss**

Colin McInnes, Adam Kamradt-Scott, Kelley Lee, Anne Roemer-Mahler, Owain David Williams and Simon Rushton: **The Transformation of Global Health Governance**

Tom Watson (editor): **Asian Perspectives on the Development of Public Relations: Other Voices**

Marketing Big Oil: Brand Lessons from the World's Largest Companies

Mark L. Robinson
President and CEO, Capitol Hill Communications, LLC, USA

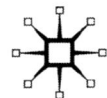

MARKETING BIG OIL

Copyright © Mark L. Robinson, 2014.

All rights reserved.

First published in 2014 by
PALGRAVE MACMILLAN®
in the United States—a division of St. Martin's Press LLC,
175 Fifth Avenue, New York, NY 10010.

Where this book is distributed in the UK, Europe and the rest of the world, this is by Palgrave Macmillan, a division of Macmillan Publishers Limited, registered in England, company number 785998, of Houndmills, Basingstoke, Hampshire RG21 6XS.

Palgrave Macmillan is the global academic imprint of the above companies and has companies and representatives throughout the world.

Palgrave® and Macmillan® are registered trademarks in the United States, the United Kingdom, Europe and other countries.

ISBN: 978-1-137-38808-7 EPUB
ISBN: 978-1-137-38807-0 PDF
ISBN: 978-1-137-38916-9 Hardback

Library of Congress Cataloging-in-Publication Data is available from the Library of Congress.

A catalogue record of the book is available from the British Library.

First edition: 2014

www.palgrave.com/pivot

DOI: 10.1057/9781137388070

This book is dedicated to the five most important people in my life:
My parents and my family.

Contents

List of Tables	viii
List of Brand Lessons	ix
Introduction	1

Part I From Standard Oil to Big Oil

1	Big Oil and the Love-Hate Relationship	6
2	The Oil Refining Era: 1863–1869	10
3	The Standard Oil Empire Reigns Supreme: 1870–1900	17
4	The End of One Oil Empire and the Beginning of Another: 1905–1911	26
5	The Arrogant and Aloof Oil Company	31

Part II Managing the Brand Crisis

6	How BP Destroyed a Corporate Brand	41
7	The Exxon Valdez: A Failure in Brand Crisis Leadership	48
8	Shell versus Greenpeace and Brent Spar	55
9	The Tarnished BP Brand: From Texas City to Price Fixing	64

10	Chevron versus Ecuador: How a Strong Brand Defends Itself	71
11	A 'Shell' Game for Investors	79

Part III Marketing Strategies and Brand Building

12	Marketing and Advertising Innovation at Mobil Oil	88
13	The Brand Disconnect between BP and 'Beyond Petroleum'	95
14	Chevron and the Evolution of Human Energy	107
15	Brand Building at Shell Oil	118

Part IV Big Oil and the Era of Consumer Engagement

16	Building Brand Loyalty: Improving the Retail Fueling Experience	130
17	Communicating with the Masses: Big Oil and Social Media	136

Part V Concluding Remarks

Index	148

List of Tables

1.1	Big oil company rankings by revenue	8
1.2	BrandZ's top 100 global brands (Big Oil)	8
11.1	Shell's reserves overstatements	83
11.2	Shell's number of shares under option at the end of the year granted to executives and other employees	85
12.1	Marketing taxonomy at the gas pump: Mobil's segmentation of consumers	92
14.1	Attitudes toward Standard Oil of California from 1944 to 1957	108
14.2	Attitudes toward Standard Oil of California before and during Arab Oil Embargo (Western states)	108
17.1	BP's tweets during the Gulf of Mexico spill	140

List of Brand Lessons

Brand Lesson No. 1	37
Brand Lesson No. 2	46
Brand Lesson No. 3	53
Brand Lesson No. 4	62
Brand Lesson No. 5	69
Brand Lesson No. 6	77
Brand Lesson No. 7	86
Brand Lesson No. 8	94
Brand Lesson No. 9	94
Brand Lesson No. 10	105
Brand Lesson No. 11	116
Brand Lesson No. 12	127
Brand Lesson No. 13	134
Brand Lesson No. 14	144

palgrave▶pivot

www.palgrave.com/pivot

Introduction: Why We Love to Hate the Oil Companies

Abstract: *The largest oil companies—Exxon Mobil, Chevron, BP and Royal Dutch Shell—have received little attention in the marketing and brand management literature, both within the academic and popular non-fiction arenas. This is primarily due to the negative image the industry had inherited ever since John D. Rockefeller's Standard Oil Trust was broken up in 1911 over its virtual monopoly of power and highly questionable business practices. Further, crises such as the Exxon Valdez oil spill in 1989 and BP's Deepwater Horizon's explosion in April 2010 and the resulting environmental and PR disasters have only served to perpetuate the negative image of the industry.*

Robinson, Mark L. *Marketing Big Oil: Brand Lessons from the World's Largest Companies.* New York: Palgrave Macmillan, 2014. DOI: 10.1057/9781137388070.0004.

The term "Big Oil" typically conjures up mental images of gigantic drilling platforms, extensive pipeline networks, oversized profits, branded retail gasoline/petrol stations, and perceived influence in cutting covert deals with unfriendly governments for the rights to explore, drill, and produce crude oil. The four largest vertically integrated oil companies referred to as Big Oil are household names within the industry: Exxon Mobil, Chevron, BP, and Royal Dutch Shell/Shell Oil (United States). The term Big Oil also refers to the seminal big oil company—Standard Oil- which was founded on January 10, 1870 as the Standard Oil Company (Ohio).

Throughout its 41-year corporate existence, Standard Oil had acquired a reputation as a company that used intimidation and other aggressive tactics to drive competitors out of business and that engaged in restraint of free trade. These unethical and often times illegal tactics led Standard Oil to become known as the largest and most financially successful oil company well into the 20th century. At one point in its corporate history, Standard Oil was also the most hated company in America. To put this in the context of today's marketing discipline, Standard Oil owned a poor corporate brand.

Operating initially from a base in Cleveland, Ohio, and in later years, New York City, many of Standard Oil's business practices were revealed first by local media and then on a national scale by the investigative journalist Ida Tarbell. Between 1904 and 1906, Tarbell issued a 21-part, scathing and highly provocative series of articles in *McClure's* magazine. Titled the "History of the Standard Oil Company," the sensationalistic story-telling held the public's attention for years and brought into sharp focus the egregious business dealings of the company. Tarbell's meticulous reporting revealed the intricacies of Standard Oil's operations first in Cleveland, then in New York, and finally, in a good part of the U.S. East Coast, leading the public to fully comprehend the outlandish methods the company used to succeed. Although many of Standard Oil's competitors were aware of how the company conducted its oil refining business, Tarbell single-handedly brought it all out in the open, making Standard Oil the most hated company in America.

In what is known to be the first example of a public relations disaster within the oil industry, Standard Oil's founder, John D. Rockefeller, Sr., failed to address Tarbell's reporting which allowed the public to create its own, highly unflattering and hated corporate image of Standard Oil. His pervasive, aloof and arrogant persona while testifying in numerous lawsuits against his company served only to fuel the public's contempt. Rather than mount public relations and advertising campaigns challenging Tarbell's accusations, Rockefeller preferred to hide behind a wall of secrecy within the confines of the company's New York City's headquarters at 26 Broadway. This elusion served only to further infuriate the public and damage the company's brand image and reputation.

Many of the lessons from Rockefeller's public relations failures have neither been learned nor assimilated into the brand management practices of today's largest publicly traded oil companies, many of them the progeny of Standard Oil. These companies have suffered the same public relations missteps as Standard Oil and have inherited many of its faulty brand management practices. Over a period of the last three decades several major events have further damaged these companies' corporate brand including BP's Deepwater Horizon spill in April 2010, Exxon's *Valdez* incident in March 1989, and Royal Dutch Shell's reserves accounting scandal in January 2004. With the benefit of hindsight, it now appears that the actual events didn't cause the most public outrage but rather how these events were handled from a PR and crisis communications perspective that has continued to impact and damage these companies' reputations and brand image.

Part I of this book provides a historical overview of the operations and business dealings of Standard Oil from its founding in 1870 to its breakup in 1911. It is from an analysis of the company's operations and founder John D. Rockefeller Sr.'s ruthless pursuit of "winning at all costs" corporate culture, where readers will gain an understanding of how the company came to be perceived as abusive and brazen in dealing with the competition. Big Oil's current poor corporate brand image has been inherited from Rockefeller's Standard Oil and has continued to be perpetuated through generations.

Part II provides a detailed look at singular events that have come to shape Big Oil's current brand image: primarily environmental disasters and the lack of leadership in handling managing a brand crisis. The discussion is not placed directly on the event itself but rather on how the company's image and reputation were impacted by poor crisis leadership. Actions that the companies could have taken to lessen the damage done to their brand image will also be described.

In Part III, the discussion turns to the marketing and advertising campaigns used by Big Oil to improve its image among various stakeholder groups. Mobil Oil's innovative use of advertorials (advertising-editorials) program and the insightful marketing research that led to the development of five types of gasoline station consumers, the Friendly Serve program, and the innovative SpeedPass device are examined and analyzed. Shell's "Let's Go" and Chevron's evolution of "Human Energy" advertising campaigns are reviewed in terms of the messages they are using to communicate to consumers and whether they have led to improvements in the corporate reputation and image of these firms.

With society in a social media revolution, Part IV explores the use of digital and social media to build consumer engagement in the companies' websites, Facebook, Twitter, and YouTube.

Part I
From Standard Oil to Big Oil

1
Big Oil and the Love-Hate Relationship

Abstract: *Big Oil is the popular name given to represent the four largest western oil companies, Exxon Mobil, BP, Chevron, and Royal Dutch Shell. These companies have had to endure a poor image with the general public for well over 100 years. They also remain popular targets for investors. This relationship is polarizing. People either love them or hate them. While Big Oil historically sits atop most annual company rankings based on revenues, these companies are ranked towards the middle or end of the pack when ranked on brand value.*

Robinson, Mark L. *Marketing Big Oil: Brand Lessons from the World's Largest Companies.* New York: Palgrave Macmillan, 2014. DOI: 10.1057/9781137388070.0006.

As far back as most Baby Boomers can remember the oil industry has had a poor brand image. As teenagers, we were exposed to the harsh reality of price shocks while learning to drive a car for the first time during the Arab oil embargo (October 1973 to March 1974). "Gasoline rationing" became part of the lexicon as motorists waited in long lines for a chance to fill up their cars with gasoline. And in December 1978, due to the political and economic upheaval in Iran, motorists around the world once again faced shortages of fuel and rising prices. In the eyes of the public, oil companies were responsible for these shortages and high oil prices. In short, Big Oil owns a poor corporate brand from which it is beginning to recover.

In a recent Gallup organization public opinion survey titled "Americans Rate Computer Industry Best, Oil and Gas Worst," the oil industry, which includes the largest of the vertically integrated oil companies—Exxon Mobil, BP, Royal Dutch Shell/Shell Oil (U.S.) and Chevron—ranked last out of 25 industries in terms of consumer popularity.[1]

Why is Big Oil so demonized? The reasons are many. First, motorists pay a price for a product where the economics of oil and gasoline supply and demand aren't well understood. When gasoline prices rise, typically in tandem with an increase in the price of oil, motorists can't miss seeing the daily reminder of high retail prices plastered on roadside signage along with the company's logo at their corner station. High gasoline prices act like a tax on motorists, taking away disposable income that might otherwise be spent on other goods and services. Second, environmental disasters such as BP's oil spill in the Gulf of Mexico in April 2010 and the Exxon *Valdez* grounding and spill in Alaska during March 1989 continue to portray the industry as owning a poor corporate image. Based on the research conducted for this book, this poor image is not entirely due to these actual events—although they do play a role—but rather how poorly the events were handled from a public relations and crisis communications perspective. The inability to strategically manage these events demonstrates that Big Oil is woefully ill-prepared to deal with environmental disasters. Third, Big Oil continues to be a money-making machine which makes these companies attractive to investors, but serves as a painful reminder for consumers of the discrepancy between large company profits and high retail prices.

Each year, *Fortune* magazine issues its Fortune Global 500 annual company rankings. For 2012, Big Oil was ranked as seen in Table 1.1.

TABLE 1.1 *Big oil company rankings by revenue*

Rank	Company	Revenues (US$mil)	Profits (US$mil)
1	Royal Dutch Shell	484,489	30,918
2	Exxon Mobil	452,926	41,060
3	BP	386,463	15,699
4	Chevron	245,621	26,895

Source: http://money.cnn.com/magazines/fortune/global500/2012/full_list

TABLE 1.2 *BrandZ's top 100 global brands (Big Oil)*

Rank	Company	Brand Value 2013 (US$mil)	Profits (US$mil)
39	Exxon Mobil	19,229	18,315
49	Royal Dutch Shell	17,678	17,781
78	BP	11,520	10,424
97	Chevron	9,036	8,599

Note: CNN Money Magazine. http://money.cnn.com/magazines/fortune/global500/2012/full_list
Source: *FT Special Report*, Global Brands: Global Top 100, *Financial Times* May 21, 2013.

Contrast the *Fortune* ranking with that of BrandZ's Top 100 Global Brands as ranked by brand value and Big Oil is seen in a different perspective as seen in Table 1.2.

In the annual Fortune ranking, Big Oil has continuously ranked at or near the top in terms of revenues and profits; in the BrandZ ranking, oil companies tend to be placed in the middle or towards the end of the pack, while companies like Apple, Google, and Microsoft ranked the highest in brand value. Fourth, consider a recent trip to the gasoline or petrol station. In terms of attractiveness, Big Oil's retail stations have been an afterthought. The stations can make for an unpleasant customer engagement experience in terms of overall appearance, lighting, and restroom facilities. At times, the benzene fumes coming from the gasoline hose can be overwhelming. The major oil companies need to place more emphasis on their retail operations as they are often the only opportunity the companies have to directly engage with their customers.

Perhaps the most important reason why oil companies operate under an umbrella of negativity is that the industry has failed to educate and connect with the public as to what value they provide to the society. That could change if Big Oil adopted many of the brand building processes and consumer engagement tools being used by today's innovative companies.

So, why do many people hate the largest oil companies? As former Shell Oil executive John Hofmeister wrote, the primary reason is a fundamentally wrong approach to corporate communication:

> Instead of being accessible to the media, many energy companies choose to buy advertising space to tell a guarded version of the truth. Instead of educating consumers on the real risks and real cost of energy, they choose to sponsor cultural and educational television programs. Instead of being on-site to respond to a crisis, they send the lawyers.[2]

So how did this love-hate relationship with Big Oil get started? It all began with one person and the company he founded: John D. Rockefeller, Sr. and the Standard Oil Company.

Notes

1. Gallup organization poll. "Americans Rate Computer Industry Best, Oil and Gas Worst." (August 16, 2012). Accessed April 11, 2014, http://www.gallup.com/poll/156713/americans-rate-computer-industry-best-oil-gas-worst.aspx
2. John Hofmeister. *Why We Hate the Oil Companies* (Palgrave Macmillan: New York, 2006), p. 7.

2
The Oil Refining Era: 1863–1869

Abstract: *The Oil Refining era began in Cleveland, Ohio where Standard Oil founder John D. Rockefeller, Sr. formed his company, not to refine gasoline or petrol as it is referred to in other parts of the world, but kerosene. Motor vehicles requiring gasoline would not appear until well into the 20th century. Refining crude oil into kerosene required access to crude oil supplies and access to a transportation method. In these early years of the oil industry, refining companies had to rely on railroads which were continually expanding throughout the United States. Rockefeller made secret deals with the railroads forming what today we would call a cartel between oil refiners and railroads.*

Robinson, Mark L. *Marketing Big Oil: Brand Lessons from the World's Largest Companies.* New York: Palgrave Macmillan, 2014. DOI: 10.1057/9781137388070.0007.

The history of the oil industry is rife with conflicts, embargoes, economically devastating price swings, and the introduction of "'big business" into the lexicon. It also provides examples of competitive intimidation, secret business arrangements, industrial espionage, and monopolistic activities. These negative images of the oil industry were based on the corporate culture of one entrepreneur and the company he founded in 1870: John D. Rockefeller, Sr. and the Standard Oil Company. As a by-product of industrial consolidation, antitrust legislation, and corporate spinoffs, the siblings of Standard Oil—now known as Big Oil—have unknowingly perpetuated the negative image of their former corporate parent some 140 years later.

The first oil strike

Within a few months following Edwin Drake's oil strike in Pennsylvania on August 28, 1859, local media began reporting on the infatuation with discoveries of black gold. On November 18, 1859, the *Cleveland Dealer* newspaper reported "the oil springs of Northern Pennsylvania were attracting considerable speculation" and that there was "quite a rush to the oleaginous locations."[1]

Every amateur geologist and prospector descended on these Pennsylvania oil fields in hopes of striking it rich. One of those individuals seeking to make his fortune in oil was John D. Rockefeller, Sr.

Rockefeller was born in Richford, in upstate New York on July 8, 1839. After spending his formative years in New York, the Rockefeller family moved to Strongsville, Ohio, a suburb of Cleveland. He entered a professional school where he perfected the practice of single- and double entry bookkeeping along with other skills that would ultimately serve him well in his future businesses. At age 16, Rockefeller landed his first job, working for Hewitt & Tuttle, commission merchants and produce shippers. From there, Rockefeller would quickly learn how to buy and sell commodities giving him valuable insight into his future business: selling refined oil products.

Rockefeller and the beginnings of oil refining

In 1863, Rockefeller, along with business partners Maurice Clark and Samuel Andrews, gained entry into the industry not by exploring, drilling, and producing oil—known as *upstream*—but by refining it into kerosene (gasoline refining would come many years later). Today, the refining division of large oil companies is called *downstream* and includes transporting finished gasoline to local area storage tanks, tanker trucks which transport gasoline from storage tanks to underground gasoline holding tanks at the local retail station, and the final sale of gasoline to consumers.

At the time he began his foray into refining, Rockefeller was a commission agent far removed from the producing oil wells. As a middleman, Rockefeller exemplified a new breed of the American entrepreneur, destined to innovate and carve out his or her own niche in the new industrial economy. These entrepreneurs were people who spent their days trading and distributing all types of commodities, mostly food products. Following several years of financial success, the team of Rockefeller, Andrews and Clark formed a new corporate entity: Andrews, Clark & Company and built the first kerosene refinery called the Excelsior Works in The Flats, Cleveland's emerging industrial area. This previously isolated and unused area would soon become an important hub in local and domestic commerce due to its proximity to the U.S. East Coast via the railroad. Locating the refinery close to the *Atlantic and Great Western Railroad* gave Rockefeller and his partners an undisputed competitive advantage over other refiners due to their lower transportation costs.

In those early days of the burgeoning oil industry, US$1,000 was the initial financial investment for owning a refinery and the employees to manage and run it. Anyone that could raise the money could soon own their very own refinery even though few business people of the day knew how to run a business much less the basic economic principles of supply and demand. Coupled with large discoveries of oil and undisciplined production levels, the industry would soon experience many of the "boom and bust" cycles that in future years would become commonplace. During the early 1860s, the overproduction of oil sent prices falling and rising to ten cents a barrel even though teamsters continued to charge US$3 to US$4 per barrel to ship it to the railroads.[2] After gaining several years of successful business experience in oil refining, Rockefeller decided to buy out his partners and in 1865, a new company, Rockefeller & Andrews was formed. This new corporate entity quickly became Cleveland's largest and most well-known refining company. A second oil refinery would soon be in operation by December 1865 called the Standard Works making Rockefeller the leading Cleveland refiner.

By 1866, fully two thirds of Cleveland kerosene was being exported to London and Paris and before long, Western Europe emerged as the most important market for American kerosene. To handle the export of kerosene, Rockefeller dispatched his brother William to New York City in 1866 to launch the firm of Rockefeller and Company.

By early 1867, Henry M. Flagler had joined with Rockefeller forming Rockefeller, Andrews & Flagler. Flagler, another entrepreneur well-known throughout the region made his living in Ohio's agricultural trade industry where he shipped most of his corn and wheat produce to Cleveland and on occasion, supplied Rockefeller with many loads of produce, which Rockefeller would dutifully later resell, acting as his agent. Several years later, Flagler would end up working for Rockefeller at Standard Oil.

On March 4, 1867, the *Cleveland Leader* newspaper made the following announcement regarding Rockefeller and his new partners:

> "This firm is one of the oldest in the refining business and their trade already a mammoth one...Their establishment is one of the largest in the United States. Among the many oil refining enterprises, this seems to be one of the most successful: its heavy capital and consummate management having kept it clear of the many shoals upon which oil refining...houses have so often been stranded."[3]

Rockefeller, the first oil industry visionary, understood the one sure way to consistently earn large profits in oil refining was to grow the business as large as possible while at the same time, making it extremely efficient through economies of scale and eliminating wasteful spending. This management skill was one of the many Rockefeller possessed that his competitors did not. One of the other outcomes of Rockefeller's growing managerial expertise was in creating new products from refining's waste processes.

The importance of the railroads

Before extensive networks of oil pipelines were built, the oil industry relied heavily on railroads for the transportation of oil. Cleveland quickly emerged as the hub of many transportation networks thus giving Rockefeller tremendous negotiating power with the railroads. In these early days in the oil industry, Cleveland was serviced by three main railroad lines that gave its inland refineries direct access to eastern ports: the *New York Central*, which ran north from New York City to Albany and then west to Buffalo, where the *Lake Shore* line paralleled Lake Erie to Cleveland; the *Erie Railroad*, which traveled across New York State to a point south of Buffalo, where the *Atlantic and Great Western* subsidiary headed down into Cleveland and the Oil Regions of Pennsylvania; and finally, the *Pennsylvania Railroad*, which went from New York and Philadelphia to Harrisburg and Pittsburgh.[4] With managerial brilliance, Rockefeller played these railroads against each other in complex schemes and secured preferential rates for his company. In some cases Rockefeller was able to secure covert and unpublished rates that allowed his company to ship crude oil to Cleveland and then refined oil to New York for only US$1.65 per barrel compared to an officially published rate of US$2.40.[5]

One business scheme devised by Rockefeller to gain these preferential freight transport rates was to supply the *Lake Shore* railroad with a guaranteed sixty carloads of refined oil on a daily basis. Although Rockefeller's refineries weren't able to refine oil at the required capacity to fulfill his ambitious pledge, he would seek "cooperation" with other Cleveland refiners so that as a group (read *cartel*), they could supply the agreed shipment. For any mode of transportation including railroads, the prospect of steady shipments of this magnitude was irresistible. What this meant was that railroads could dispatch trains composed solely of oil tank cars instead of a random assortment of freight cars that had to pick up various products along the way. By consolidating small oil shipments into one large shipment allowing for continual and uniform shipments, the railroads became more efficient, reducing the average round-trip time of their trains to New York from thirty days to ten and operate a fleet of 600 cars instead of 1,800.[6]

By acquiescing to Rockefeller's sound business logic of transport efficiency, the railroads had a vested interest in the creation of an oil refining arrangement (read *monopoly*) that could lower their costs and increase their profits. Rockefeller would soon grant them their wish and as time went on, the railroads developed a stake in the growth of any big business with which they could make similar arrangements. At the same time these special arrangements were being formulated, small and unprofitable refining plants were unceremoniously weeded out from competition, or were acquired altogether by Rockefeller and eventually closed.

Price crash

By 1869, oil production in the Cleveland region was rampant; so much oil was being produced that prices rose, and then quickly collapsed in rollercoaster fashion. Tracing oil prices back ten years earlier revealed some interesting insights. The price of oil stood at about 10 cents per barrel, and then in 1861 it rebounded to ten dollars. By 1862, the price fluctuated between 10 cents and US$2.25 per barrel, averaging US$1.50 a barrel for the year. Eventually, the average price of a barrel of oil was US$3.50 in 1863, US$8.00 in 1864, US$4.00 in 1866, US$2.80 in 1867, US$5.80 in 1869, US$4.20 in 1871, and less than US$2.00 in 1873.[7] Rockefeller watched from the sidelines as oil was being recklessly overproduced. He came to one logical conclusion which would have a long-standing impact on the nascent oil industry: a market with unrestrained production would, in the long run, be unprofitable, and could be detrimental to his own enterprise. The disastrous results produced by those refiners — volatile oil prices —who did not understand the basic economic principles of supply and demand, were undisciplined, and were out to make a quick buck, led Rockefeller to distrust the principles of the free market. By his own calculations, Rockefeller estimated that over three-fourths of petroleum refiners worked at an economic loss.[8] His visionary plan would soon force a major industrial correction.

The grand plan takes shape

As the 1860s drew to a close, Rockefeller began to craft his own plan to reign in the overproduction of oil. His solution was to control not one part of the market, but rather, the entire market—oil production, transportation and storage, and refining—so that supply and demand for refined oil products reached equilibrium. Rockefeller was a business trailblazer who meticulously planned and executed his cooperation strategy without the privilege of modern-day economic textbooks or management consultants. His plan sought to reign in excess production, stabilize prices, and further consolidate the industry through the creation of a giant monopolistic cartel.

To execute his plan, Rockefeller needed large sums of money, first to create economies of scale that would allow his company to become the low cost producer and enough cash reserves to endure economic downturns, which had become all too common. Flagler's idea was to supplement their capital without relinquishing control. The solution was to incorporate the business which would enable them to sell shares to select outside investors. By the late 1860s, many states had passed legislation permitting companies to incorporate. The problem for Rockefeller and Flagler was that these firms were not permitted to own property outside their state of incorporation. Thus, companies had to restrict themselves to operating within the confines of their headquartered state. On January 10, 1870, the partnership of Rockefeller, Andrews and Flagler was abolished in favor of a newly created corporate entity: the Standard Oil Company (Ohio). The new name—an early example of oil industry corporate and product branding—would emphasize the quality of its kerosene product at a time when inferior products were being sold. Already a mini-monopoly within the state of Ohio and neighboring states, Standard Oil not only controlled 10 percent of American petroleum refining, but an oil barrel-making plant, warehouses, shipping facilities, and a fleet of railroad tank cars.[9] By acquiring these individual pieces of the industry and integrating them into one holistic unit, Standard Oil became the first vertically integrated oil company in history and laid the foundation for today's Big Oil company structure.

Notes

1 Ron Chernow. *Titan: The Life of John D. Rockefeller* (Random House: New York, 1998), p.76.
2 Ibid., p. 81.
3 Ibid., p. 108.
4 Ibid., p. 111.
5 Ibid., p. 113.
6 Ibid., p. 113.
7 Harold F. Williamson and Arnold R. Daum. *The American Petroleum Industry: The Age of Illumination, 1859–1899.* (Evanston, IL: Northwestern University Press, 1959), p. 118.
8 Chernow, *Titan,* p. 130.
9 Ibid., p. 132.

3
The Standard Oil Empire Reigns Supreme: 1870–1900

Abstract: *Once Standard Oil was formed in 1870, the company had already acquired 10 percent of the oil refining market in the United States and throughout the 1870s to 1890s, Standard Oil was the largest oil company in the world. While its profits were large, its brand image and corporate reputation were poor. To the public and many local U.S. governments and politicians, Standard Oil was the most hated company in America. The company acquired other refining companies for pennies on the dollar and those that were not acquired were forcibly put out of business.*

Robinson, Mark L. *Marketing Big Oil: Brand Lessons from the World's Largest Companies.* New York: Palgrave Macmillan, 2014. DOI: 10.1057/9781137388070.0008.

On January 10, 1870, the partnership of Rockefeller, Andrews and Flagler was abolished and was replaced by the Standard Oil Company (Ohio). According to one source, the new company began operations with the following shareholders: John D. Rockefeller Sr. (30 percent), William Rockefeller (13.34 percent), Henry Flagler and Samuel Andrews (each holding identical shares of 16.67 percent), Stephen Harkness (13.34 percent), and O.B. Jennings (brother-in-law of William Rockefeller, 10 percent).[1] At the time of its founding, the company's market share of the United States' oil refining business was estimated at 10 percent.

Rockefeller quickly moved to consolidate the entire, and what he deemed to be chaotic, oil refining business in Cleveland, which was becoming the main producing center in the United States. In one lightning strike within two short months, February and March 1872, he acquired 22 of the 26 refining companies in what would come to be known as the *Cleveland Massacre*.[2] During one 48-hour period in early March of the same year, Rockefeller bought six refineries. He continued this aggressive acquisition spree nationwide for several more years until Standard Oil controlled virtually all American oil refining. Between the stock market crash—*Black Thursday*—on September 18, 1873, and autumn 1874, Rockefeller continued his campaign to bring the industry under the operational control of Standard Oil. With quick and decisive strokes of managerial brilliance, Rockefeller provided the catalyst for a chain reaction that spelled the end of independent oil refining not only in Cleveland, but also in neighboring centers in Pittsburgh and Philadelphia, Pennsylvania. Of the 22 Pittsburgh refiners still remaining after the buying spree began, only one was still independent two years later.[3]

How was Rockefeller able to purchase these competing refineries with little negotiation in price? Using his expert accounting skills which were perfected over years of business dealings, he paid, on average, only a quarter of their original construction costs. In one case, Rockefeller paid US$45,000 for a refinery that could have easily fetched US$75,000 at local auctions.[4] But in the majority of cases, Rockefeller paid next to nothing for goodwill—the intangible value in an ongoing company such as its corporate reputation, logos, or client lists. Part of Rockefeller's grand plan was to take over these competing plants not to continue to operate them as in most mergers and acquisitions transactions, but to shut them down entirely, thereby eliminating excess capacity and potential competition.

The South Improvement Company scheme

While Rockefeller continued to acquire and eliminate competing refineries, his plan of gaining favorable terms with the railroads remained a work in progress. Under the terms of a new proposed cooperative agreement (*cartel*) with a Rockefeller-led group of refiners called the South Improvement Company (SIC), Rockefeller was able to persuade the railroads to sharply raise published freight rates for all refiners, except for refiners in the SIC. These refiners would receive secret and substantial discounts—up to 50 percent off crude and other refined-oil-shipments—whereby their competitive advantage over rival companies would exponentially increase. In the most anti-competitive part of this pact, the SIC members would also receive "drawbacks," which were payments on shipments made by competing refiners. In today's marketing lexicon, we would call these drawbacks rebates; that is the railroads would give the SIC members rebates for every barrel they shipped for *other* refiners. In one example using shipments from oil fields in western Pennsylvania to Cleveland, Standard Oil would receive a 40-cent discount on every barrel it shipped, plus another 40 cents for every barrel shipped to Cleveland by its competitors! One Rockefeller biographer went so far as to call the drawback scheme "an instrument of competitive cruelty unparalleled in industry."[5] Once this pact became publicly known, it was the first major instance of negative publicity surrounding the oil industry and severely damaged Standard's brand image. Through another provision in the alliance, Standard Oil management and other SIC refiners would receive detailed information about oil shipped by their competitors, creating an unethical opportunity to underprice them.[6]

Once the SIC scheme became more widespread, the *Oil City Derrick* trade magazine began to publish a daily list of SIC members, including Rockefeller, on its front page. Each day, its readers would see a new inflammatory caption depicting Rockefeller and Standard Oil as a large reptile, squeezing the life out of the competition. One such caption read, "Behold the Anaconda in all his hideous deformity."[7] And when the media discovered where Rockefeller lived and sought interviews, Rockefeller turned them away empty-handed after which he instituted a new corporate policy: instructing all Standard Oil personnel to say nothing to reporters. By remaining silent in the face of criticism, Rockefeller thought the public would see him as confident and secure in his integrity when to many, he seemed guilty, arrogant, and aloof. And by not creating his own public relations and advertising offense, Rockefeller allowed the media to create its own unflattering image. This would foreshadow the day when oil companies created internal public relations departments accompanied by marketing and advertising campaigns to deflect the negative media stories.

As Standard Oil's dominance in oil refining grew, so too did its entrenchment within the railroad industry. Standard Oil was like a tick on the animal hide of the railroads, constantly sucking favors out of them until it had full operational control. The company profited immensely as oil transport in barrels gave way to a new innovation in oil transport: railroad tank cars. In 1874, Standard Oil began a fundraising campaign to find a way to economically build oil-tank cars, which it would then lease back to the railroads for a special mileage allowance.[8] As the sole owner of a majority of the *Erie* and *New York Central* tank cars, Standard Oil's leading position grew unbeatable so much so that the company could dictate operational and pricing policies. At a moment's notice, the company could either bring railroad to its knees or force a complete shutdown by threatening to withdraw its tank car fleet. Moreover, Standard Oil was able to force the railroads into granting Standard Oil special arrangements for tank cars not used by the small refiners who still preferred to ship oil by barrel. In one example, railroads would levy a charge for the return of empty barrels, while tank cars traveled free on the return trip from the East Coast to Midwest refineries. Tank-car clients would also receive the exact same leakage allowance received by barrel shippers, even though the tank cars didn't leak. This special arrangement allowed Standard Oil to carry 62 gallons free in every tank car.[9]

By the mid-1870s, Rockefeller and Standard Oil had heavily invested not only in the railroad companies themselves, but also in tank cars. Although pipelines were in their infancy, Rockefeller's railroad investments were being threatened by improvements in pipeline technology. Up to now, pipelines could pump oil only short distances, about five miles, from the production wellhead to the railroads, but when the *Pennsylvania Railroad* expanded into pipelines via its freight subsidiary, the Empire Transportation Company, Rockefeller had to acknowledge this new competitive disruption. This recent expansion into pipelines presaged a time when pipelines would span great distances and could potentially supplant railroads altogether.

Not to be out-maneuvered by his rivals, Rockefeller quickly developed a countermove by constructing his own pipeline network under the auspices of the newly created American Transfer Company. Standard Oil also acquired a one-third ownership in Vandergrift and Forman, another pipeline company. Through his association with Jacob J. Vandergrift, a steamboat captain who had earlier merged his refining interests with Standard Oil, Rockefeller and Vandergrift formed the core of a new venture, the United Pipelines. The new company secretly disguised its organizational ownership structure and activities to be free of Standard Oil control and influence. By awarding small stakes in United Pipelines to William H. Vanderbilt of the *New York Central* and Amasa Stone of the *Lake Shore* railroads, Rockefeller tightened his grip over friendly railroads. This afforded him the opportunity to extract maximum advantage from both the railroads and pipelines so long as these two means of oil transport coexisted in the oil business. By the time the owners of the first pipeline systems established a pool to set rates and allocate quotas among competing networks during the summer of 1874, Rockefeller's pipelines owned an impressive 36 percent share of the oil transport market.[10]

Let the lawsuits begin

Just before the end of the 1870s, lawsuits challenging Standard Oil's industrial supremacy began to be filed. One of the first lawsuits filed occurred on April 29, 1879, when a grand jury in Clarion County, Pennsylvania, indicted nine officials from Standard Oil—including Rockefeller, Flagler, and newly appointed John Archbold—and charged them with conspiracy to monopolize the oil business, extorting railroad rebates, and manipulating prices to cripple its competitors. Those Standard Oil officials, who resided in Pennsylvania, were arrested and released on bail while those such as Rockefeller were able to evade prosecution. But even as he was able to stay one step ahead of the law, Rockefeller genuinely feared that the negative publicity being generated in the news media would lead to more lawsuits being filed and more pressure being brought by local politicians. On December 24, 1879, Standard Oil publicly rescinded many of its abusive policies against its rivals. In one example, Standard Oil renounced the use of covert rebates and consented to publicly post its freight rates. Further, the United Pipe Lines would no longer engage in discrimination among its many pipeline shippers. In return for this gesture, both criminal and civil cases against Standard Oil in Pennsylvania were withdrawn. In time, it would be revealed that these actions were merely a superficial way to settle these cases, while the abusive and competitive behavior continued.

The trust agreement

By the early 1880s, Rockefeller controlled 90 percent of U.S. refineries and pipelines, owned the vast majority of tank cars used for rail transport, and the entire production of high-grade railroad lubricants and kerosene. He was even able to manufacture his own oil barrels cheaper than others could. But as his empire grew, it became fragmented and unmanageable. As the end of the 1880s drew near, the United States had no law allowing for federal incorporation, a situation that, in Rockefeller's estimation, made it so that a corporation created in one state made it against the law to hold assets in another state.

The solution that Rockefeller's brilliant management team devised was the *Trust*, which was becoming a standard way of doing business in the late nineteenth century. By creating the Standard Oil Trust, Rockefeller created a highly centralized planning structure with enormous market power. Under its state charter, Standard Oil of Ohio couldn't own companies outside the state, so it developed an ingenious solution: it assigned three midlevel managers to serve as trustees who held stock in a score of subsidiaries outside of Ohio. When these managers received dividends, they distributed them to the 37 investors of Standard of Ohio *as individuals*, in amounts proportionate to their ownership stakes in the parent company. This newly built structure permitted Rockefeller to swear under oath during legal proceedings that Standard Oil of Ohio didn't own property outside of Ohio, even though it controlled most of the pipelines and refineries in Pennsylvania, New York, New Jersey, and Maryland. From a legal standpoint, the trustees owned these properties.[11]

The 1879 trust agreement, as bold and innovative as it was for its day, lasted only three years. When the state of Pennsylvania attempted in 1881 to tax the property of Standard of Ohio, Rockefeller feared that other states might copy this precedent and tax it as well. The management team once again devised a plan without the need for management consultants in which they could unite *intrastate* firms into an *interstate* giant. The first step was to create a separate Standard Oil company in each state where it held properties. As a result, Standard Oil of New York (a precursor to Mobil Oil) was formed on August 1, 1882, with William Rockefeller as president. Four days later, Rockefeller himself became president of the new Standard Oil of New Jersey (a precursor to Exxon, now Exxon Mobil). This strategic move was designed specifically to prevent each state from taxing Standard Oil property located outside the state. The new Standard Oil trust agreement was dated January 2, 1882 and most surprisingly, the public knew nothing of the new structure. Imagine a company that in those days was worth an unheard of sum of US$70 million, which controlled 90 percent of American refineries and pipelines, and the general public knew next to nothing about it.

Standard Oil's marketing districts

Rockefeller's plan to solidify and control the refining of crude oil into kerosene demonstrated the lengths he would go to control a market. To bring kerosene to local markets, he divided the United States into eleven marketing territories. Each marketing territory sales manager was expected to maintain the share of their local market: the benchmark was to achieve as close to 85 percent as was possible which led them to not only aggressively compete against rival firms but also with each other. One way they achieved local market dominance was to routinely undersell their rivals (underselling rival firms was one of Standard Oil's most potent weapons owning to its efficient distribution system and operating as the low cost producer). Whenever a Standard Oil marketing territory sales manager learned of a new competitor entering its market, several actions were taken. First, the company would sell its product at cost while raising the price in other markets to compensate. They would then keep prices low enough so as not to raise suspicion or lose control of the market. Second, Standard Oil marketing managers habitually engaged in foot surveillance. Using this covert practice, Standard's marketing managers would secretly follow a competitor's tank wagon from town to town and, if necessary, undersell them. This action fostered the development of an extensive competitive intelligence network whereby territory managers would forward their field reports to Standard Oil's headquarters at 26 Broadway in New York City. This information would then be compiled and analyzed and placed into actionable strategic and tactical intelligence.

The firm also made use of what today would be called industrial espionage. One Cleveland oil refiner was horrified to discover that Standard

Oil paid his own bookkeeper US$25 per month for information on their own refined oil shipments!

A surprise discovery

A shocking example of Standard Oil's domination of the oil industry was discovered quite by accident almost a decade later in the 1890s. As recounted by Rockefeller biographer Ron Chernow, while perusing the books in a local shop, then Ohio Attorney General David K. Watson came across a book with a title that caught his eye; *Trusts: The Recent Combinations in Trade*. The book included a written copy of the Standard Oil Trust deed. After reviewing the trust agreement he was surprised to see that for the last seven years, Standard Oil had been violating its state charter by having transferred control of the company to out-of-state trustees in New York! Representing his constituents in Ohio, Watson filed a petition against Standard Oil (Ohio) in the state supreme court in May 1890 seeking nothing more than a complete breakup of the company. On March 2, 1892, Watson won his victory when the Ohio Supreme Court ruled that Standard Oil (Ohio) was, in fact, controlled by trustees at the 26 Broadway headquarters and had to renounce the trust agreement. On March 10, 1892, the company announced that the trust would be dissolved. Once again, Rockefeller and "the Standard," as it was often referred to, managed to escape from the clutches of total annihilation at the hands of the law. With changes to New Jersey's incorporation law having recently been enacted, the Standard Oil Company of New Jersey took on a unique status as a transformed company: it bought whole or parts of blocks of stock in the other Standard Oil companies so it could function as both an operating company *and* a holding company. No wonder then that Standard Oil of New Jersey (now Exxon Mobil) was the most financially successful and largest of the parent spinoffs.

Notes

1 Corporation for Public Broadcasting (PBS). American Experience: The Rockefellers. http://www.pbs.org/wgbh/americanexperience/features/biography/rockefellers-john/.
2 Ibid.

3. Ibid.
4. Ibid., p. 147.
5. John T. Flynn. *Men of Wealth: The Story of Twelve Significant Fortunes From the Renaissance to the Present Day* (Simon and Schuster: New York, 1941), p. 444.
6. Chernow, *Titan*, p. 136.
7. Hidegarde Dolson. *The Great Oildorado* (Random House: New York, 1959), p. 265.
8. Chernow, *Titan*, p. 137.
9. Ibid., p. 170.
10. Ibid., p. 172.
11. Chernow pp. 224–225.

4
The End of One Oil Empire and the Beginning of Another: 1905–1911

Abstract: *In 1911, the U.S. Supreme Court dissolved the Standard Oil Trust which for the past 40 years had become a virtual monopoly not only in oil refining into kerosene, but also in manufacturing its own oil barrels, railroad tank cars, and oil pipelines. Even though the company was dissolved, its corporate offspring, Exxon, Mobil, Chevron, and Amoco, became Fortune 500 companies. Their profitability grew due to the invention of the gasoline engine, gasoline powered cars, the expansion of the U.S. highway system, and the development of the retail gasoline station.*

Robinson, Mark L. *Marketing Big Oil: Brand Lessons from the World's Largest Companies.* New York: Palgrave Macmillan, 2014. DOI: 10.1057/9781137388070.0009.

By the early 1890s, and despite its financial success, the image of Standard Oil was becoming tarnished. The monopoly had suffered several and significant legal setbacks and, along with Ida Tarbell's investigative reporting, found itself on the negative side of public opinion. In 1892, the Ohio Supreme Court ruled that Standard's official trust arrangement was, in fact, an Ohio company that was controlling out-of-state companies, and was therefore breaking the law. Although court documents revealed that Standard's operational business policies had actually resulted in higher quality oil products and lower consumer prices, the trust arrangement was nevertheless still considered undesirable because of its negative impact on competition.[1] Rather than dissolve their ownership in the company as the court had instructed, the company's trustees totally ignored the court's ruling and merely rearranged their properties. The Standard Oil Trust—now some 84 companies strong—was officially dissolved and eventually reorganized so that by 1899 the same people still controlled the same company, only this time, it was from New Jersey under the newly formed company Standard Oil of New Jersey.[2] But as these legal issues were increasing, a number of other cataclysmic events were about to unfold that foreshadowed the end of the Standard Oil monopoly altogether.

The curtain begins to fall

In September 1901, newly elected U.S. President McKinley was assassinated making Vice President Theodore Roosevelt president. Roosevelt's stated contempt for trusts would eventually target Rockefeller and Standard Oil. As part of his federal government policy, Roosevelt distinguished between good trusts and bad trusts. Good trusts provided benefit to society, bringing products to market that benefitted consumers. On the other hand, bad trusts, such as Standard Oil, restrained free trade, bullied competitors, and engaged in abusive business practices.

What led Roosevelt's decision to target Standard Oil came primarily from the 21 installments Ida Tarbell had authored in *McClure's* magazine. Running monthly from 1904 through 1906, Tarbell's "The History of the Standard Oil Company" described the inner workings of the company in minute detail, revealing the entirety of the company's actions, many of which would today be illegal. Tarbell's investigative reporting captured the attention of the American public for those two years, making the fiercely private Rockefeller the center of attention. Further, anti-monopoly sentiment in the United States had also been building for many years as the mainstream media reported on virtually every move Rockefeller and Standard Oil made. The public's furor had been growing for years and not even Standard Oil could advertise or buy its way out of its poor corporate brand image.

By early 1901, oil strikes were made in other parts of the United States, most importantly, Spindletop in Texas, which were followed by new and sizeable discoveries in California, Oklahoma, Kansas, and Illinois. Standard Oil's executives were especially surprised by discoveries in Texas where they had been for years confidently declaring there was no oil to be found. As the story goes, oil drillers in Beaumont went prospecting for water, but were disappointed when they found oil. At the time, farmers in the Lone Star State needed water more than oil, and for a while they traded a barrel of oil for a barrel of water. Not even the massive physical size and financial muscle of Standard Oil could bring these newfound supplies to market and manage it, much less compete nationwide against new and far-flung rivals.

The federal government lawsuit

By the summer of 1907 lawsuits filed against Standard Oil had been rising exponentially. By then there had been no less than seven federal and six state lawsuits brought. Much of the information uncovered during these legal proceedings revealed that, despite the lawsuits and judgments by the courts against it, Standard Oil had still been engaging in illegal behavior. An Interstate Commerce Committee report issued in January 1907 revealed that the company was still secretly accepting rebates from numerous railroads, spying on competitors, setting up bogus subsidiaries, and engaging in predatory pricing. Not surprisingly, these activities had been ruled illegal for many years.

Of all of the lawsuits filed against it, the biggest legal case of all was to be the federal case against Standard Oil filed under the Sherman Antitrust Act. Hearings in the federal lawsuit lasted from 1907 to 1909 and included over 400 witnesses testifying. In November 1909, the U.S. District Court in St. Louis ruled unanimously that Standard Oil of New Jersey had conspired to restrain interstate trade and the court recommended that the corporation be dissolved. Standard appealed to the U.S.

Supreme Court, but on May 15, 1911, it agreed that Standard was guilty as charged, and ordered the company's dissolution within six months.

The convergence of cars and gasoline refining

This momentous legal decision might have foreshadowed the end of Standard Oil had it not been for several innovations throughout the remaining years of the nineteenth century and the early years of the next.

During the early 1880s, future automobile manufacturer Gotlieb Daimler strapped light gasoline engines onto 2-wheeled bicycles and tricycles. In 1886, a future Daimler colleague, Karl Benz patented a three-wheeled automobile-type structure with a single cylinder engine. Henry Ford, who would later play a major role in automobile manufacturing via the assembly line, had beta-tested a two-cylinder engine for a streetcar. Brothers Frank and Charles Duryea mass-produced 13 two-cylinder runabouts in Springfield, Massachusetts; this was the first time a car company had mass-produced several cars from a standardized model. The innovations that would make Rockefeller even wealthier in retirement than in business were the convergence of the mass production of cars, the retail gasoline or filling station, and the expansion of the U.S. highway system. These innovations would propel Rockefeller and the prodigy of Standard Oil into the stratosphere of Corporate America.

Standard Oil of New Jersey eventually became Exxon and then Exxon Mobil—and never lost its lead as the largest publicly traded oil company in the world. Standard Oil of New Jersey was more generally known as Esso, an acronym for Eastern Seaboard Standard Oil, but the other Standard Oil companies objected. To end the confusion, Esso was renamed as Exxon in 1972.

Standard Oil of New York (Socony) soon acquired Magnolia Petroleum and then General Petroleum, which made it a force in refining the crude, while Vacuum became a major operator of gasoline service stations. Socony and Vacuum achieved many synergies, which made the future merger a reality. In 1931 they united to form Socony-Vacuum, still remembered by oil historians for the flying red horse trademark. In 1955 the company's name was officially changed to Socony Mobil, and in 1966 was shortened to Mobil.

Other Standard Oil spinoffs became large corporations in their own right. There was also Standard Oil of California which became Chevron; Standard Oil of Ohio, which became SOHIO, the American arm of BP; Standard Oil of Indiana, which became Amoco; Continental Oil which later became Conoco; and Atlantic, which became ARCO and then eventually, Sun. These companies in turn experienced rapid growth during the 1950s and 1960s due mainly to the rise in automobile purchases, the rapid growth of the retail gasoline market, and the expansion of the interstate highway system to become some of the world's largest companies. Still, for years following the breakup, the Standard companies cooperated with one another even though legally they were independent. The government knew what was happening, but did little to halt the practice, believing that to interfere would cause harm to the industry and economy as a whole.

During the merger wave of the 1990s and early 2000s, these spinoffs of Standard Oil then merged to form the following:

- Exxon and Mobil became Exxon Mobil Corporation
- British Petroleum purchased both Atlantic Richfield (ARCO) and Amoco and kept its name as British Petroleum (now simply BP)
- Chevron and Texaco later merged and became Chevron

On the other side of the Atlantic Ocean, and not part of the original Standard Oil family of companies, Shell Transport and Trading and Royal Dutch Shell combined to form the other piece of Big Oil (after the company restructured, the official name became Royal Dutch Shell PLC). Shell Oil is the U.S. subsidiary of Royal Dutch Shell.

Notes

1 Chernow, *Titan*, p. 332.
2 B. Bringhurst. *Antitrust and the Oil Monopoly: The Standard Oil Cases, 1890–1911.* (Greenwood: Westport, CT, 1979), pp. 20, 32–33.

5
The Arrogant and Aloof Oil Company

Abstract: *John D. Rockefeller, Sr. was notoriously aloof and arrogant in his business dealings with competing oil refiners, railroads, and local government officials. During his many court testimonies, his behavior was aloof and arrogant. After the breakup of the Standard Oil Company, executives from the new companies—Exxon, Mobil, Chevron, and others—were themselves equally as arrogant and aloof as Rockefeller. Throughout the 1960s and 1970s, these new oil executives took it upon themselves to use the same strategies, as did Rockefeller. By relying on arrogant and aloof behavior, all of these executives helped to create and perpetuate a poor corporate brand image and reputation.*

Robinson, Mark L. *Marketing Big Oil: Brand Lessons from the World's Largest Companies.* New York: Palgrave Macmillan, 2014. DOI: 10.1057/9781137388070.0010.

During the period 1895–1910, it was becoming readily apparent that a majority of Americans held unfavorable attitudes toward all large businesses with Standard Oil serving as the archetype.[1] For decades, Standard Oil had virtually disintegrated all competition by utilizing both legitimate and highly questionable business practices. In a short period of time, Standard Oil had risen from only one of ten Cleveland refiners to practically the one remaining refining company in the United States.[2] A commonly used estimate is that by 1880 the company controlled about 90 percent oil refining and oil transport in the United States.

Although Rockefeller and his management team were able to prevent most competition from impacting Standard Oil's business interests well into the 1890s, politics, especially antitrust sentiment, were on the horizon. There had always been minor skirmishes with lawsuits before, and both Rockefeller and Standard Oil have been able to skillfully evade the majority of them. But it wasn't until July 2, 1890, when U.S. President Benjamin Harrison signed the Sherman Antitrust Act into law that Rockefeller began to comprehend the full impact that politics and lawsuits would have on both him and Standard Oil. Years before Ida Tarbell's sensationalistic journalism would appear in *McClure's* magazine and the Sherman Antitrust Act became law, cracks in Standard Oil's armor appeared beginning with the March 1881 issue of the *Atlantic Monthly*. The "Story of a Great Monopoly" was the first serious exposé of the Standard Oil Trust in a prestigious, mass circulation magazine. The article introduced Rockefeller to a nationwide audience and put Standard Oil and the issue of antitrust first and foremost on the political agenda.

Many oil historians deem Tarbell's series as the coup de grâce for both Rockefeller and Standard Oil. Not only was the series well researched and sensationalistic, the duration kept the American public on the edge of their seats for years. Both Rockefeller and Standard Oil were never out of the limelight for those two years.

The series officially consisted of two parts, the first with nine installments and the second with eight.[3] These two parts lasted from November 1902 to October 1904. Tarbell then published four additional articles on Standard Oil in 1905 and two on Rockefeller himself, bringing the entire series to 21 articles published in four years, with the great majority being highly unflattering. Imagine, in comparison, a four year exposé on the secret business operations of Exxon Mobil, Royal Dutch Shell/Shell Oil (U.S.), Chevron, and BP by the *Wall Street Journal* or the *Financial Times of London*.

The publication of Tarbell's series in book form further increased Standard's exposure to an already hostile public. She supplemented the material from the *McClure's* series with meticulous research including numerous documents including newspaper articles from the *New York Times,* court transcripts, letters, and lists showing the complete organizational structure of the parent company and its numerous subsidiaries. The inclusion of these documents served to bolster the book's appearance of objectivity. Not surprisingly, the book gained national attention. Awareness of Standard's nefarious operations from both the media and the public were only heightened by Tarbell's publications and the failure of Rockefeller and Standard Oil to respond to her allegations only added more fuel to an already burning fire.

Aloofness

Throughout its corporate existence, Standard Oil rarely responded to negative criticism of its operations, business practices, and management style. What responses it did make failed to counter the accusations made by the media and the general population. By not mounting aggressive public relations responses, the company allowed these stakeholders to create their own negative brand image of the company.

The first manifestation of Standard's arrogance and self-importance was its aloofness, a type of corporate behavior first described by Benoit.[4] The company continually disregarded public opinion and rarely responded to its critics. Since 1870, Standard Oil's policy was never to respond to criticism or negative press for fear that any response might lead it to inadvertently disclose proprietary or sensitive information. Because Tarbell's charges were predominately historical in nature, Standard Oil never felt it necessary to respond. The company demonstrated a continuing willingness to ignore these accusations, even when there was compelling evidence of illegal behavior. This inaction implied a lack of concern of wrongdoing. In one example, when members of his management team pressed him to respond to Tarbell's accusations, Rockefeller replied, "Gentlemen, we must not become entangled in controversies. If she is right we will not gain anything by answering, and if she is wrong, time will vindicate us."[5]

In another instance, Standard Oil demonstrated its aloof behavior by deciding not to list itself on the stock exchange. Listing on a stock exchange typically means that a company is required to issue quarterly

DOI: 10.1057/9781137388070.0010

and annual financial reports, along with other useful investor information. In this case, Standard elected to maintain absolute secrecy by not publicly trading its stock.

Even in the face of continuous negative media coverage, Standard Oil never deemed it necessary to develop a public relations function within the company. The company finally hired a press agent in 1906 but by then, it was too late; the damage to Standard Oil's reputation had been done and was irreversible.

Bolstering

A second strategy implemented by the company is that of bolstering. Bolstering is an example of a tactic used by both Standard Oil and Exxon during its mishandling of the *Valdez* brand management crisis in 1989. When using the bolstering technique, companies focus on their own greatness and importance to the industry.

When in a letter to private shareholders following President Roosevelt's Bureau of Corporations' attack on Standard for continuing to receive rebates for shipments of oil from the railroads, it did defend itself, but determined that a defense was not necessary: "it is surely not within the limits of fairness for the Bureau of Corporations to cast aspersions upon a great corporation."[6] It concluded that Standard's success "is not traceable to illegal or reprehensible methods, but to its economic and elaborate industrial organization, covering as it does every detail of transportation, manufacture, and administration."[7] Rockefeller reinforced this idea publicly in an interview after the breakup of the company by saying that Standard "has been one of the greatest, if not the greatest, of upbuilders we have ever had in this country—or in any country."[8]

In another example of its bolstering activities, the company oftentimes referred to itself as "the Standard." In a 1906 letter to the company's shareholders, company secretary C.M. Pratt wrote of "competitors of the Standard."[9] This usage was not only observed in the company's internal documents, but also in how the media referred to it. Writers also referred to the company in this same way; news stories of the day often quoted Standard officials using the same term, "the Standard."[10] This usage reinforced the public perception of the trust as monopolistic and further implied that there was only one "Standard" and that it was exalted above all other corporations, even non-oil companies.[11]

The aloof oil executive—Rockefeller testifies in court

As more lawsuits against Standard Oil began to be filed in various state jurisdictions and tried in numerous courts of law, it became commonplace for Rockefeller to be seen and heard testifying in court proceedings. Up until now, Rockefeller was able to hide from both the public and the media, but it was becoming clear that he had to begin to show his face in public and face his accusers. The strategies and tactics Rockefeller and Standard Oil's legal counsel used to evade prosecutor's questions were not only ahead of their time, but they are still in use today by Big Oil's executives.

In one lawsuit, Rockefeller gave identical replies to 30 consecutive questions asked of him, declaring each time, "I refuse to answer on the advice of counsel."[12] In another example, prosecutors spent five hours grilling Rockefeller on the various charges being brought against the company. In what had become common testimony strategy tactics, Rockefeller would speak softly in the lowest volume possible and provide as little information as possible (one might conclude that a conscious decision was made by Rockefeller to reveal as little information as possible so as to protect the company's proprietary data and trade secrets, but we may never know the true answer).

In a 1907 case brought against Standard Oil of Indiana, the accusation was that the company had taken illegal rebates from the *Chicago* and *Alton* railroads. The shipments in question had passed between Whiting, Indiana and East Saint Louis, Illinois, *after* the Elkins Act outlawed such rebates. The presiding figure in the Chicago courtroom that day was Judge Kenesaw Mountain Landis. By this time, Rockefeller was "a virtuoso of evasive testimony."[13] As summarized by one Rockefeller biographer, Judge Landis began questioning Rockefeller as follows:[14]

"Mr. Rockefeller, what is the business of the so-called Standard Oil Company of New Jersey?" "I believe, your Honor..." Rockefeller responded, and then appeared to lose his train of thought. He paused, shifting position in the witness chair, and then made a second attempt at an answer. "I believe, your Honor..." Once again he appeared to be confused. After another few attempts, Rockefeller responded by saying, "I believe your Honor, they operate an oil refinery in New Jersey." To all of Landis' follow on questions, Rockefeller would respond in the same slow, drawn out, methodical and disconnected style, making what little testimony he had given, virtually useless.[15] Following additional testimony, Judge Landis delivered his verdict: a fine against Standard Oil of Indiana that dwarfed any other in American corporate history up until that time: a fine of US$29.24 million dollars (US$457 million in 1996 dollars).[16] This maximum penalty was based on the following formula: US$20,000 for each of 1,462 carloads of oil cited in the indictment. In typical boisterous style, Rockefeller responded, "Judge Landis will be dead a long time before this fine is paid"[17] (this utterance by Rockefeller shows how determined both Standard Oil was and Big Oil companies are to *not* pay any legal fine brought against it by a court of law, seeking to wait the court and the prosecution out as long as possible before making any payment. Exxon's legal strategy in the 1989 *Valdez* episode was to continually appeal a courts' ruling thereby causing a delay in making any payment until several decades later. Chevron has implemented this same strategy in the case against citizens in Ecuador over oil spilled.(It has yet to pay any fine levied against it by a court of law). What was even worse for the prosecution was that during the next year a federal appeals court not only revoked the fine against Standard Oil but severely reprimanded Landis for considering each carload as a separate offense. In a retrial, Standard Oil was found *not guilty*. President Theodore Roosevelt, a hater of trusts, especially Standard Oil, was furious.

The present day oil executive

Just as Rockefeller had been aloof and arrogant during his testimony, so too are today's Big Oil executives. During the Arab Oil embargo of October 1973 to March 1974, U.S. Senators and Congressmen held hearings on the energy crisis from two perspectives. The first dealt with the crisis itself and the role, if any, played by the oil companies, and the second from the viewpoint of consumers. Despite mainstream media reporting on the oil companies' obscene profits, the large majority of consumers were left to endure long lines at the gas pump, or in some cases, gasoline stations had completely run out of the fuel and were closed altogether.

Of the many hearings during the energy crisis, none garnered the most media attention as those held by the Senate Permanent Subcommittee on Investigations.[18] The committee, unable to resist the temptation of playing to the harsh impact on consumers and the high profits made by the oil companies, scored a major political coup by lining up the executives of the seven largest companies at one table and having them testify under oath. Under the microscopic lens of television lights and cameras, oil company executives were subjected to question after question regarding their operations and financial performance. While many executives had education and experience in the petroleum sciences such as geology, petroleum and chemical engineering, they were ill prepared to deal with a brand crisis of this magnitude (and many are still ill prepared today). During the hearings, they came across as inept, insulated, self-important, and out of touch with the common man and woman.

In terms of the oil industries' "obscene profits," the committee wanted to know how they were possible in light of the embargo. Historically, the profits of the largest oil companies had been close to flat for the five years ending in 1972, despite the explosive growth in demand. Profits then rose from US$6.9 billion in 1972 to US$11.7 billion in 1973, and grew exponentially to US$16.4 billion in 1974.[19] Much of the increase came from the companies' foreign operations. As oil-exporting countries pushed prices up by lowering supply, the oil companies got a free ride in terms of increases on the non-U.S. equity oil they still owned. The value and the market prices of their American oil reserves also increased.[20] Further, they had bought oil at lower prices before the increases, kept it in inventory, and then profited when the oil was finally sold at a higher price.

As Rockefeller and other members of his management team were able to avoid even the most direct and specific prosecutorial questions, today's Big Oil executives have mastered the art of evasive testimony. In a 2005 oversight committee testimony for the U.S. House Energy and Commerce Committee, members of the committee attempted to question BP's then former manager in charge of pipeline corrosion prevention, Richard Woollam. At the time, Woollam was on paid leave and even refused to answer the committee's questions, invoking his Fifth Amendment rights against self-incrimination.[21]

Brand lesson no. 1

Despite all of the negative publicity, the many lawsuits surrounding Rockefeller and Standard Oil, and the questionable business practices he set in motion, the current operational structure of today's oil industry, upstream (exploration and production); midstream (transportation and storage); and downstream (marketing, refining, and retail) was created.

Rockefeller possessed the visionary insight that three related industries should be consolidated and the managerial discipline of wringing excess cost out of each operational area. Further, Rockefeller's management team created the first competitive intelligence system, collecting, analyzing, and disseminating (via the newly invented telegraph), information on Standard Oil's competitors comprising the areas of market expansion, supply, demand, and pricing. With no formal training in the oil industry or the field of management consulting, Rockefeller was a fastidious practitioner of detail in accounting and managerial oversight. Although the original Standard Oil Trust was formerly dissolved in 1911, the offspring of Standard Oil have become the world's largest companies and some of the most profitable in the world, more profitable, in fact, than even Rockefeller himself could have possibly imagined. But despite all of Standard Oil's business innovations and financial success, Rockefeller's decision to use immoral and a highly contentious operating culture had unknowingly inflicted long-lasting damage to the nascent oil industry's image.

Notes

1. L. Galabos. *The Public Image of Big Business in America, 1880–1940* (The Johns Hopkins University Press: Baltimore, 1975), p. 120.
2. Josh Boyd. "The Rhetoric of Arrogance: The Public Relations Response of the Standard Oil Trust," *Public Relations Review,* 27 (2001), pp. 163–178.
3. Ibid., p. 168.
4. W.L. Benoit. *Accounts, Excuses, and Apologies: A Theory of Image Restoration Strategies* (State University of New York Press: Albany, 1995).
5. Chernow, *Titan*, p. 442.
6. C.M. Pratt, secretary, Letter to shareholders from the Directors of Standard Oil Company, (May 16, 1906), Standard Oil collection, University of Wyoming American Heritage Center, Laramie, Wyoming, p. 9.
7. Ibid. p. 10.
8. J.D. Rockefeller, Interview with William O. Inglis 1917–1920 (Meckler Publishing: Westport, CT, 1984) p. 17.
9. See footnote 24, pp. 3, 6, 8.
10. Boyd. "The Rhetoric of Arrogance" pp. 163–178.
11. Ibid.
12. Chernow, *Titan*, p. 540.
13. Ibid., p. 540.

14 Ibid., p. 214.
15 Ibid., p. 542.
16 Ibid., p. 514.
17 Ibid., p. 542.
18 Daniel Yergin. *The Prize: The Epic Quest for Oil, Money & Power* (The Free Press: New York, 1991), p. 656.
19 Ibid., p. 658.
20 Ibid., p. 658.
21 *Oil Spill Intelligence Report*, 29(38) (September 14, 2006), p. 1.

Part II
Managing the Brand Crisis

6
How BP Destroyed a Corporate Brand

Abstract: *In 2000, BP changed its corporate image from a traditional oil company to one that marketed a more caring environmental image. The company suffered blows to this new image as a result of several major catastrophes, the 2005 explosion at the Texas City, TX refinery and the Deepwater Horizon spill in April 2010. Although the environmental impact has and will be debated for years, the event serves as a case study of how a properly planned and executed PR and crisis communications plan can help reduce the damage done to the corporate brand.*

Robinson, Mark L. *Marketing Big Oil: Brand Lessons from the World's Largest Companies.* New York: Palgrave Macmillan, 2014. DOI: 10.1057/9781137388070.0012.

"I want my life back."[1] These words, spoken by BP's former CEO Tony Hayward a little more than a month after the April 20, 2010 rupture of the Macondo oil well, will forever be etched in the minds of the public. The disaster was nothing like the oil industry had ever seen, much less envisioned: high pressure methane gas and other liquid drilling byproducts spewed from the well onto the Deepwater Horizon drilling rig causing an explosion and fire that killed eleven crewmen.[2] Although the environmental impact has and will be debated for years, the event serves as a case study of how a properly planned and executed PR and crisis communications plan can help reduce the damage done to the corporate brand.

In the weeks and months following the event, BP suffered multiple blows to its brand. The first and most immediate was the impact on its stock price, which plummeted to a 14-year low, slashing the company's pre-crisis market value by half.[3] BP eventually suspended its dividend payments to shareholders. The company also divested billions of dollars in assets to ensure sufficient financial resources were available to fund compensation claims and to complete the cleanup effort.

According to YouGov's Brandweek Buzz Report, BP's brand rating also suffered. The report is a weekly consumer perception survey that collects and analyzes data on the most talked about brands based on marketing buzz. The scores are based on weighing positive and negative perceptions of a brand. A score of +100 is positive, while a score of −100 indicates negativity. A rating of zero means the score is neutral. In YouGov's April 12, 2010 report, immediately prior to the spill, BP had a +58.4 recommend score. In its June 4, 2010 ranking, just eight weeks after the disaster, YouGov reported BP's score had dropped to −37.4.[4] The BP spill impacted other oil company brands as well. Shell's brand score quickly dropped from +42.7 to +32.2, while Mobil fell from +37.5 to +28.1 within one weeks' time.[5] BP's Gulf of Mexico (GoM) spill, perhaps dredging up painful memories of the Exxon *Valdez* incident, caused Exxon's rating to drop from +15 to −8.9.[6]

The crisis takes shape

As BP was attempting to manage the crisis, the company made efforts to express regret and to take full responsibility for the damage caused by the explosion and the spill. In an attempt to create the impression that the company was sincere and remorseful through numerous media statements, its executives, in actuality, damaged the BP brand through a series of public relations missteps. These missteps were only exacerbated by consumers' use of social media.

Faulty leakage reporting

Given BP's technical expertise when it comes to drilling for oil, it's hard to find a reason why the company was unable to accurately assess the amount of oil leaking from the well even weeks after the incident. Two days after the explosion, the company sent a remote-controlled robot to the well site to assess the level of damage. BP reported that no oil was leaking and it appeared for a short time the company had averted a major catastrophe. But on April 24, that initial assessment changed dramatically. The company then reported that close to 1,000 barrels per day (bpd) were leaking from the oil well with that estimate being increased by the United States Coast Guard some three days later to 5,000 bpd. By the end of the month, the world's attention was transfixed on the BP Deepwater Horizon crisis, both in traditional media, social media, and live video streaming from the ocean floor. As several attempts were being made to cap the well, both BP and the U.S. federal government raised the estimated amount of oil leaking from the well. The U.S. Geological Survey in mid-May 2010 raised its estimate from 12,000 to 19,000 bpd, while the federal government again raised its estimate to 60,000 bpd. The actual number of barrels spilled remains a mystery to this day as BP continues to downplay the size of the spill. The U.S. Department of Justice estimates a total of 4.2 million barrels while BP claims it as 2.45 million.[7]

The CEO speaks

As the crisis grew, CEO Tony Hayward appeared to grow weary and frustrated with BP's attempts to cap the well. On May 30, 2010, Hayward delivered an apology during an interview with a reporter. His comments seemed to suggest that his time was better spent elsewhere rather than managing a crisis of this magnitude. During the interview Hayward stated: "We're sorry. We're sorry for the massive disruption it's caused to their lives. And there's no one who wants this thing over more than I do. I'd like my life back."[8] While this comment was meant to be sincere, the public interpreted his message very differently. During the interview Hayward was observed feeling frustrated and irritated and appeared to be disassociating himself from BP. In the eyes of the public, Hayward was not separate from BP; he *was* BP. And when he testified during U.S. Congressional hearings a month later Hayward stonewalled his way through seven hours of intensive grilling on the decisions made in the days and hours leading up to the spill. Throughout his testimony, Hayward refused to answer the questions posed to him. Many of his responses ranged from *"I was not involved in that decision, so it's impossible for me to answer that question,"* to *"I'm afraid I can't recall that,"* or, simply, *"I don't know."*[9]

A clash of cultures

One of the critical elements that lend credibility to corporations in any brand crisis is being able to identify with the victims. While this action can be achieved through interviews and prepared media statements, the ways in which senior executives relate to the victims is of paramount importance in flawlessly executing the crisis communications playbook. When Hayward appeared on the beach shortly after the crisis began, even his attire suggested an "us versus them" mentality. Appearing in an expensive business suit while expressing frustration over the time commitment necessary to manage the crisis conflicted with the blue-collar working class of the citizens of the Gulf region, many of who rely on the local seafood industry for their livelihoods. Hayward's public display of his elite class mentality was confirmed three weeks later when the media published photographs showing him with his son at a yacht race on the Isle of Wight.[10] Hayward's thick British accent likely exacerbated the negative connotations of his resentful statements because it identified him and BP as foreign, and a non-American who might not care about the United States coastline.[11]

BP suffered a further brand setback when both Hayward and board chairman Carl-Henric Svanberg appeared at the White House alongside President Barack Obama to announce the development of the US$20 billion compensation fund. Following the announcement, Svanberg attempted to assure the American public that the company appreciated and would respond to the plight of those whose livelihood had been harmed, and he attempted to express an appreciation for President Obama's concern for the people along the Gulf Coast who had been put out of work. With his thick Swedish accent, Svanberg said, "We care about the small people. I hear comments sometimes that large oil companies are greedy or don't care. But that is not the case at BP. We care about the small people."[12] Based on Svanberg's poor choice of words, BP replaced Hayward shortly thereafter with employees from the Gulf Coast region in the company's marketing and advertising campaigns. Hayward was eventually replaced in the CEO role by Robert Dudley, who was a native Mississippian.

The BP brand four years later

When oil company brands encounter a crisis, the basic principles of recovery and image restoration are to acknowledge the error, express remorse for the victims and their families, and let all stakeholders know of the remedial actions taken by the company. As of this writing, the restoration of the BP brand remains a work in progress.[13]

The role of brand in gasoline/petrol retailing

A brand is used as a measure to understand the impact of a product's image in driving demand within a product category.[14] Up until the BP brand crisis, convenience of location and low prices were dominant drivers for gasoline/petrol retailers. The Gulf of Mexico disaster changed all that with U.S. consumers going out of their way to avoid their local BP gas station (during the Exxon *Valdez* brand crisis of 1989, many consumers cut up their gasoline credit cards and boycotted Exxon gas stations in reaction to the Alaskan spill). So while BP may remain somewhat insulated from permanent damage to its commercial operations due to the world's continued consumption of oil, consumers and investors remain conflicted over the brand.

Media lessons

Tony Hayward's poor leadership during the crisis led directly to his ouster. Because the personification of the ultimate responsibility rested with him, as it does with the CEO of any company, media training should be compulsory after witnessing Hayward's difficulties in front of the glare of the media's 24-hour appetite for news. Hayward may have wanted to demonstrate a natural, caring, and remorseful persona, but his comments and actions didn't translate well into the arena of global media.

BP and greenwashing

BP was one of the first brands to truly encounter the wrath of the social media sphere in the wake of the Gulf of Mexico spill as the public used the opportunity to proclaim their distaste for the company. The brand that had claimed it wanted to be "beyond petroleum" appeared to be communicating an empty promise.

Many other oil company brands appear to be deluding themselves into believing they have partial responsibility for the delivery of their marketing messages. Oil company brand managers are falsely telling themselves that it is acceptable for the inside world of a company to have a limited relation to the way it wants to be perceived and seen by its customers. In the world of social media—primarily Facebook, Twitter, and YouTube—it is easy for customers to openly express anger at their brands.

Brand lesson no. 2

What could BP have done differently that might have reduced the damage done to its brand? One action the company could have taken was to publicly show its remorse for the victims and their families. Hayward should have attended the funerals of the 11 workers killed by the explosion. Executives from the health care products company Johnson & Johnson are seen by many as a model for companies impacted by a brand crisis to follow. In 1982, when Tylenol Extra Strength tablets were discovered to be tainted with cyanide, executives attended the funerals of the seven victims and were filmed openly weeping at the gravesides. That is the type of human compassion BP lacked.

At the outset of the crisis, it was correct that Hayward should be the public face of BP. However, to truly identify with the victims, BP realized too late that by using local company employees as spokespersons the company would likely have been able to suppress much of the negativity surrounding the spill. Moreover, both Hayward and Svanberg used politically incorrect language to describe both the situation and the victims.

Notes

1. See BP CEO Tony Hayward. "I'd Like My Life Back," *YouTube* (May 30, 2010), accessed October 24, 2013, http://www.youtube.com/watch?v=MTdKa9eWNFw.
2. Campbell Robertson. "Search Continues after Oil Rig Blast," *New York Times* (April 21, 2010).
3. Ayesha Rascoe and Tom Doggett, *BP Shares Plunge as U.S. Threatens New Penalties, Reuters,* accessed June 9, 2010, http://www.reuters.com/article/2010/06/09/us-oil-spill-idUSTRE6573FD20100609.
4. *Brandweek.* "Oil Companies Rebound, Except BP" (June 4, 2010), http://www.adweek.com/news/advertising-branding/oil-companies-rebound-except-bp-102527.
5. Ibid.
6. Ibid.
7. "BP Still Trying to Minimize the Size of the Deepwater Horizon Spill," *The Times-Picayune* (October 9, 2013). http://www.nola.com/opinions/index.ssf/2013/10/bp_is_still_trying_to_minimize.html.
8. See endnote 1.
9. Gail Russell Chaddock. "BP Oil Spill: Tony Hayward's Stonewalling Approach Before Congress," *The Christian Science Monitor* (June 17, 2010), p. 12.
10. Erin O'Hara O'Connor. "Organizational Apologies: BP as a Case Study," *Vanderbilt Law Review* (November 2011), pp. 1958–1991.
11. Ibid.
12. For a video clip of these statements, see Associated Press, *BP Chief.* "We Care About the Small People," *YouTube* (June 16, 2010), http://www.youtube.com/watch?v=th3LtLx0IEM.
13. "A Shrunken Giant," The *Economist* (February 8, 2014), p. 61.
14. Graham Hales. "The BP Brand One Year Later: Branding Lessons from the Disaster," *Interbrand* (2010).

7
The Exxon *Valdez*: A Failure in Brand Crisis Leadership

> **Abstract:** *Following the grounding of the oil tanker Exxon Valdez in 1989, the company and its chief executive, Lawrence Rawl, greatly mishandled the public relations response to the incident. The company's executives attempted to use several of William Benoit's image restoration strategies, such as minimization and bolstering in an attempt to deflect the negative publicity surrounding the event.*

Robinson, Mark L. *Marketing Big Oil: Brand Lessons from the World's Largest Companies.* New York: Palgrave Macmillan, 2014. DOI: 10.1057/9781137388070.0013.

The Exxon Valdez incident has been analyzed and well documented since its occurrence more than 25 years ago. Up until the BP Deepwater Horizon incident of April 2010, Exxon's misfortune was often cited as the "largest oil spill ever in North America."[1] The facts remain indisputable.

In the early morning of Friday, March 24, 1989, the oil tanker Exxon Valdez ran aground on a shallow reef off Bligh Island in Prince William Sound, Alaska.[2] The amount of oil discharged from the damaged tanker numbered some 240,000 barrels of oil (over 10 million gallons).[3] The grounding of the tanker created not only severe environmental damage to the spill area but also curtailed the livelihood of local small businesses, and reduced Exxon's short-term profitability. Poor crisis communications further damaged Exxon's brand.

Damage to the Exxon brand at the time of the incident, was measured by the many stakeholders impacted by the accident; the company and all of its employees; the residents and fishermen from Valdez, Alaska; the state of Alaska; the U.S. Coast Guard; the U.S. government; and consumers of Exxon's products. By mid-April 1989, Exxon reported that 6,000 cardholders had mailed back their cards to the company.[4] By May 1, the number had risen to 10,000.[5]

In the weeks and months following the spill, Exxon's crisis communications plan began to unravel to the point where it was operating in reactive mode. In a series of media stories, the public first learned of the initial spill followed by its impact on the environment and area wildlife. These stories were soon followed by allegations that the Valdez's tanker captain, Joseph Hazelwood, had made several poor judgments. First, and against Exxon's corporate policy, Hazelwood ordered the tanker be put on automatic pilot in confined water. He also set a potentially risky course, and left the ship in the hands of an inexperienced third mate. It was then reported that Hazelwood had been intoxicated during the time leading up to, and after, the tanker had run aground. It didn't help his or Exxon's case when Hazelwood disappeared from public view. In July 1989, several months after Hazelwood had become front page news, a new media story—and new brand challenge—surfaced concerning destruction of evidence: erasure of taped recordings of telephone conversations about the spill and the cleanup effort by Exxon.

Image challenges faced by Exxon

While the event was an environmental disaster and the poor spill response created an image of a corporate giant not knowing how to manage the brand crisis, Exxon attempted to implement several corporate image restoration strategies as described by Benoit: shifting the blame, minimization, bolstering, and corrective action.[6]

Shifting the blame

As soon as the U.S. government had released the results of Hazelwood's blood test, Exxon took action. Hazelwood was fired after federal investigators reported that blood tests, taken more than ten hours after the tanker struck the reef, showed that his alcohol level was well above the U.S. Coast Guard maximum for anyone in charge of such a large vessel.[7]

Exxon's action may have had some positive impact on the company's image. The company identified and had eliminated the cause of the problem. Thus, the guilty party (Hazelwood) was punished by being fired from his job and rendered unable to commit the same tragic act again. However, it was then reported that Exxon "admitted it knew the captain had gone through an alcohol detox program, but still had put him in command of the largest tanker in its fleet."[8] This action, while not completely exonerating Exxon, might have helped the company's image, but given the fact that Exxon had previously known about his alcohol problems and still selected him to captain meant that the company still shared a majority of the blame.

After Exxon had attempted to place the blame on the spill directly on Hazelwood's shoulders, it then launched a media attack on the U.S. Coast Guard and the state of Alaska for delaying the use of chemical dispersants.[9] Here, Exxon tried to shift the blame from itself onto other organizations. It was later revealed that chemical dispersants would not be effective because the wind and sea were too calm.[10]

Minimization

Exxon's first media statement indicated that it "did not expect major environmental damage as a result of the spill" was an example of the minimization image restoration strategy.[11] The use of this strategy also failed when, during BP's Deepwater Horizon incident, CEO Tony Hayward attempted to minimize the spill by remarking, "The Gulf of Mexico is a very big ocean. The volume of oil and dispersant we are putting into it is tiny in relation to the total water volume."[12]

Even as the effects of the spill became clearer, Exxon continued to use the strategy of minimization. David Ranseur, spokesperson for Alaska governor's office, declared "Exxon has gone out of its way to minimize the effects of the spill, by understating the number of animals killed and miles of beaches affected."[13] More specifically, "on May 19, when Alaska retrieved corpses of tens of thousands of sea birds, hundreds of otters, and dozens of bald eagles, an Exxon official told National Public Radio that the company counted just 300 birds and 70 otters."[14] Nor was this the only implementation of this defensive strategy. At that years' Exxon stockholder's meeting, "Mr. Rawl continued to talk of the 'good news' of a record salmon catch in Alaska, even after a local fisherman pointed out that those salmon were not from the spill area."[15]

Bolstering

Bolstering is an image restoration strategy that stresses good behavior of the company under attack during a brand crisis. Exxon attempted to use the bolstering strategy by publishing a full-page "Open Letter to the Public" from Chairman Lawrence Rawl in several newspapers. The advertisement contained a brief statement, about 170 words, that makes four points, three of which appear to be designed to bolster the image of the company. First, he praised Exxon's actions by stating: "Exxon has moved swiftly and competently to minimize the effect this oil will have on the environment, fish, and other wildlife. Furthermore...we have already committed several hundred people to work on the cleanup." Second, he asserts that, "since March 24, the accident has been receiving our full attention and will continue to do so." Third, he expresses sorrow and states "We at Exxon are especially sympathetic to the residents of Valdez and the people of the State of Alaska." In each of these utterances, Rawl attempts to bolster Exxon's image by stressing the positive behavior of the company during the event.

Taken individually and collectively, these media statements failed to bolster Exxon's image for several reasons. Consider Exxon's claim that the company's cleanup efforts were swift and competent. Evidence reaching the public from non-company sources dramatically refutes these assertions, revealing in actuality that its actions were neither swift nor competent. A full emergency crew did not arrive on the scene for over 14 hours. Despite plans to contain oil spills within five hours, this one was not contained for 35 hours. After five days, less than 2 percent (4,000 of 240,000 barrels) of the spilled oil had been collected. The available evidence directly contradicts Rawl's attempts to bolster the company's image.

Rawl's "Open Letter to the Public" also stated the incident received the full attention of the company. However, Rawl did not publicly appear to take an interest in the spill for some time. "It was not until March 30, six days after the accident, that Mr. Rawl made his first comments about the incident. He finally went to Alaska on April 24, three weeks after the event.[16] Rawl's worst error was to send a succession of lower-ranking executives to Alaska to deal with the spill instead of taking charge himself. Thus, while apparently designed to bolster the company's image, the newspaper advertising campaign seriously backfired.

Predicting the event

Could the grounding of the Exxon *Valdez* tanker have been predicted? Possibly, as any means of oil transport could fall victim to the laws of probability. Could Exxon's poor handling of the Valdez accident have been predicted from a crisis communications perspective? The answer to this question is "yes." Other analyses of the *Valdez* incident have previously overlooked the leadership decisions made by CEO Lawrence Rawl during the eight years leading up to the spill.

Untimely decisions

What many consumers don't know was that the 1989 spill in Prince William Sound had been anticipated as far back as 1981. That year the Alyeska Pipeline Service Company, an eight-company consortium of which Exxon is still a member, which operates the port of Valdez and is responsible for mounting a defense following spills, was required by law to submit a contingency plan spelling out how it would respond. Although it had submitted its contingency plan on time, one of the key fatal decisions Rawl made was to disband a 20-member emergency team prepared for a round-the-clock response to oil spills in Valdez Harbor and Prince William Sound. The reason given for this decision was that a full team wasn't required and would be a waste of human and financial resources.[17]

Who's in charge?

To exacerbate the situation, Rawl had also created a decentralized decision-making process whose unintended consequence would create severe flow-of-information problems during a crisis. This decision, based on a cost-cutting rationale, forced decision-making down to operating levels to where it would be the least effective.[18] The decentralized plan also reduced Rawl's appearances before the media, securities analysts, and even business organizations. This action resulted in a more inward-looking Exxon rather than one that should have looked outward to include society, the environment, and many more stakeholders other than its shareholders.

In July 1989, roughly three months after the *Valdez* tanker ran aground, an internal memo issued by the Exxon manager in Valdez indicated that a company pullout of cleanup activities by mid-September 1989 was not negotiable. This memo infuriated Congressional leaders in Washington, DC. At a hearing held shortly after the pullout memo, then Exxon President William Stevens was forced to reaffirm Exxon's commitment not to abandon the cleanup, revising his Valdez manager's July memo. Steven's would later describe the internal memo as "inappropriate for outside readers" and "widely misinterpreted."[19]

Brand lesson no. 3

In retrospect, the CEO is the public face of the company, and needs to be shown taking charge of the brand crisis. Rawl had realized too late that part of Exxon's top executive information problem during the *Valdez* crisis was be attributed to its decentralized structure, which put the entire responsibility for organizing public relations on a one-man show in Houston.[20] Other experts suggest, "Exxon wasn't prepared for the microscopic glare of publicity because it had been increasingly inward-looking in recent years."[21] Thus it could be suggested that a corporate policy of pushing decision-making down to operating levels, information control at the local level, and an inward-looking perspective would seem to indicate that a CEO could not be bothered in managing a brand crisis.

Notes

1   James P. Hill. "After Bhopal and Valdez: Rethinking the Semantics of Public Affairs," *MAJB* 5 (2), pp. 3–14.

2. David E. Williams and Glenda Treadaway. "Exxon and the Valdez Accident: A Failure in Crisis Communication," *Communications Studies*, 43 (Spring 1992), pp. 56–64.
3. Ibid.
4. J. Holusha. "Chairman Defends Exxon's Efforts to Clean Up Oil," *New York Times* (April 19, 1989), p. 21.
5. P. Shabecoff. "Valdez Townspeople Angered as Oil Slick Continues to Expand off Alaska," *New York Times* (April 1, 1989), p. 8.
6. Benoit. *Accounts, Excuses, and Apologies*.
7. J. Mathews and C. Peterson. "Oil Tanker Captain Fired after Failing Alcohol Test: Exxon Blames Government for Cleanup Delay," *Washington Post* (March 31, 1989), p. A1.
8. K. Wells and C. McCoy. "Out of Control: How Unpreparedness Turned the Alaska Spill Into Ecological Debacle," *Wall Street Journal* (April 3, 1989), pp. A1, 4.
9. Mathews and Peterson. p. A1.
10. M.W. Brown. "Oil on Surface Covers Deeper Threat," *New York Times* (March 31, 1989), p. A12.
11. P. Shabecoff. "Exxon Vessel Hits Reef, Fouling Water That Is Rich in Marine Life," *New York Times* (March 25, 1989), p. 42.
12. Katherine Faulkner. "BP Boss Implies Oil Slick Is Nothing More Than a Drop in the Ocean," *The Mail Online* (May 14, 2010).
13. R.W. Baker. "Critics Fault Exxon's PR Campaign," *Christian Science Monitor* (June 14, 1989), p. 8.
14. Ibid.
15. Ibid.
16. John Holusha. "Exxon's Public Relations Problem," *New York Times* (April 21, 1989), pp. D1, 4.
17. K. Schneider. "Under Oil's Powerful Spell, Alaska Was Off Guard," *New York Times* (April 2, 1989), p. A1.
18. See endnote 16, p. 4.
19. W. Stevens. Statement of W.D. Stevens, President of Exxon Company, USA, delivered before the U.S. House Water, Power and Offshore Energy Resources. *Subcommittee of the House Committee on Interior and Insular Affairs* (July 28, 1989).
20. A. Sullivan and A. Bennett. "Critics Fault Chief Executive of Exxon on Handling of Recent Alaskan Spill," *Wall Street Journal* (March 31, 1989), p. B1.
21. See endnote 16.

8
Shell versus Greenpeace and Brent Spar

Abstract: *During the mid-1990s, Shell's decision to dispose of the Brent Spar oil storage platform at sea, drew angry responses from the environmental organization Greenpeace and many European consumers. Shell attempted to counteract much of the negative media coverage while Greenpeace achieved a public relations victory against a major oil company.*

Robinson, Mark L. *Marketing Big Oil: Brand Lessons from the World's Largest Companies.* New York: Palgrave Macmillan, 2014. DOI: 10.1057/9781137388070.0014.

The North Sea has remained one of the most prodigious oil-producing regions in the world. More than forty billion barrels of oil equivalent (boe) have already been produced from the North Sea beginning in the 1970s and estimates of remaining oil reserves are between twelve billion and twenty-four billion, based on what may be viable to extract at current oil prices.[1] As with any mature producing area, offshore oil drilling facilities abound, from drilling platforms to oil storage and tanker loading facilities. Currently, there are about four hundred offshore petroleum platforms in the North Sea, half of which are in the UK sector.

The removal of oil platforms is governed by a variety of international regulatory policies. According to the guidelines of the International Maritime Organization (IMO), any installation in shallow waters must be completely removed and dismantled on land. However, a substantial portion of the current installations—about 50 in UK waters—are in deeper waters, and if approved by local governments, could be disposed of at sea. That is, oil platforms in deep waters could be towed out to sea to a disposal site and sunk. According to British legal interpretations of international conventions and guidelines as well as UK legislations, oil operators have to submit their preferred disposal option, called the Best Practical Environmental Option (BPEO), for government approval. Each case for disposal is individually considered on its merits. If the platforms are to be disposed of at sea, any remains have to be left at least 55 meters below the surface. All proposals have to be well documented and must include a rigorous review of the options being considered. The costs of abandonment are to be borne by the field licensee. In some instances some 40 to 60 percent is tax deductible.

Royal Dutch Shell, one of the largest vertically integrated oil companies, was confronted with a brand crisis involving the disposal of an offshore structure in 1994. At the time, the Brent Spar was a cylindrical buoy, measuring 470 feet high and weighing about 15,000 tons. During the 15 years between 1976—when it was first commissioned—and 1991, the Brent Spar was used as an oil storage facility and tanker-loading buoy for the Brent Field, which along with Brent Spar was 50 percent owned by Esso AG, a unit of Exxon Mobil. In 1991 an internally commissioned study concluded that the necessary refurbishing of the facility was economically unjustifiable. Brent Spar was thus officially decommissioned in September 1991. Shell UK, one of the operating companies of Royal Dutch Shell, considered several dismantling options. These options were evaluated according to several factors including engineering complexity, risk to health and safety of the disposal workforce, the environmental impact, cost, and acceptability to British authorities and other official stakeholders, the latter including regional governmental bodies such as the Scottish National Heritage and the Joint Nature Conservancy Committee, as well as "legitimate users of the sea" (as specified in the 1987 Petroleum Act), mainly fishermen's associations and British Telecom International which had placed undersea cables.[2]

During Shell's internal discussions, two options survived the initial screening process: horizontal onshore dismantling and deepwater disposal where the buoy would remain intact. The first option would consist of rotating the buoy to the horizontal, transporting it to shore, and then finally, onshore dismantling. The latter option involved towing the structure to a deepwater disposal site in the Northeast Atlantic and sinking the platform. The study commissioned by Shell UK concluded that deepwater disposal dominated on the rationale of engineering complexity, risk to health and safety of the workforce, and finally, cost (£11 million versus £46 million for onshore dismantling). Both alternatives were acceptable to the other parties consulted.

With respect to possible environmental impacts, the study concluded that both options were equally balanced. Whereas the environmental impact was expected to be minimal for both options, horizontal dismantling—due to its considerably higher engineering complexity—would involve an increased potential for mishaps that, if one were to occur in shallow water, could have a damaging impact on other users of the sea. In addition, a research team at the University of Aberdeen recommended deep-sea disposal. Consequently, Shell UK proposed deepwater disposal in its BPEO to the British Department of Energy, the relevant regulatory agency. In mid-February 1995 the British Energy minister, Tim Eggart, announced that Shell's BPEO was accepted. European governments that were affected were informed about the decision and were given three months to protest the decision. Although some of the European governments, including Germany, were generally critical of deep-sea disposal, no government officially protested, and so Shell UK scheduled the towing of the Brent Spar platform to the disposal site in the North Atlantic for mid-June.

Enter Greenpeace

Founded in 1971, Greenpeace has been the most vocal and active of the non-governmental organization (NGO) environmental groups. At the height of its influence, it had about three million contributors worldwide and a budget of about US$150 million. Offices were located in 32 countries with an estimated full-time staff of roughly 1,500 employees. Greenpeace also owned four ships, a helicopter, and modern communications equipment. It could also draw on a wide network encompassing thousands of volunteers. Greenpeace's offices were fairly independent but coordinated their decisions through Greenpeace International, located in Amsterdam. Greenpeace strategically located strongholds in Germany, the Netherlands, and the United States, where hundreds of large companies were headquartered.[3]

One of the largest and most active Greenpeace locales was in Germany. This branch of Greenpeace had about 120 full-time employees, a budget close to US$55 million, and could rely on potentially six hundred thousand volunteers.[4] Its German headquarters and the North Sea logistic centers—perfectly situated near a major oil-producing region—were located in Hamburg. Greenpeace enjoyed high acceptance and popularity among the German people and had frequently captured center stage through spectacular actions which served to increase donations that reached record amounts in 1994. Greenpeace Germany alone contributed over 40 percent of the total budget of Greenpeace International.

One of Greenpeace's principal strategies was to attract the public's attention through high-profile confrontations with major corporations, which were, in turn, covered by Greenpeace's photo and videographers. According to Steve D'Esposito, an American executive director of Greenpeace International, the group's strategy was to "keep it simple."[5] The strategy encompassed two steps. The first step is to raise awareness of the issue. Step two is to push the world toward possible solutions, using the most egregious methods. The overall strategy can be ultimately summed up as: getting in the way. Confrontation is critical to get media coverage and to reach the public.[6]

The Brent Spar protests take hold

After being informed of Shell's decision concerning Brent Spar in summer 1994, Greenpeace commissioned its own policy study to consider the arguments for deep-sea disposal. Its study was at odds with Shell's in that it concluded that total removal and not deep-sea disposal should be adopted as the BPEO, especially from the viewpoint of the environment.[7] By March, Greenpeace had devised a bold plan to board the Brent Spar. To win public support for its plan, Greenpeace acquired satellite communications and video equipment.

On April 30, 1995, 14 Greenpeace protestors from the UK, the Netherlands, and Germany landed on the Brent Spar by boat. They were later joined by a group of nine journalists who, along with Greenpeace, filmed the incident and broadcast it live by satellite. After a three-week occupation, the group was expelled by Shell. Although the UK media gave little coverage to the Greenpeace campaign, German television broadcasted extensive footage of water-soaked activists. Harald Zindler, then head of the branch of Greenpeace Germany, who organized the infamous Brent Spar landing, was quoted as saying, "We were very happy when Shell decided to clear the platform. The action portrayed Shell as an unresponsive and inconsiderate big business."[8] In response to the great outpouring of media coverage, expressions of outrage and protest in Germany and the Netherlands grew. Members of all German political parties and the German minister of the environment, Angela Merkel, condemned Shell's decision to dump the rig into the deep sea.[9] On May 22, the worker representatives on Shell Germany's supervisory board expressed "concern and outrage" at Shell's decision to "turn the sea into a trash pit."[10]

Under pressure, executives of Shell Germany met with Jochen Lorfelder of Greenpeace, who argued that close to 100 percent of German motorists surveyed would participate in a boycott. He told Shell "in the four weeks it would take to tow the Brent Spar to its dumping site, Greenpeace would make life a nightmare for Shell."[11] In response, the chairman of Shell Germany explained that Shell UK's studies indicated that deep-sea disposal was the best alternative for the environment. Lorfelder answered, "But Joe Six-Pack won't understand your technical details. All he knows is that if he dumps his can in a lake, he gets fined. So he can't understand how Shell can do this."[12]

On June 7, Greenpeace activists once again landed on the Brent Spar but were soon expelled. The next day, the Fourth International North Sea Conference began in Esbjerg, Denmark. One of the topics under discussion was the disposal of petroleum facilities. Germany introduced a proposal that would rule out any disposal at sea. Norway, France, and the UK, however, blocked the proposal.[13] In the meantime, calls for an informal boycott of Shell by German motorists were mounting. Proponents included members of all German political parties, unions, motorists' associations, the Protestant Church, and the former chief justice of the German Constitutional Court, Ernst Benda.

In its integrated marketing communications strategy, Greenpeace successfully appealed to the German enthusiasm for recycling. In their homes many Germans separated garbage into bags for metal, glass, paper, chemicals, plastic, and organic waste. Harald Zindler pointed out the emotional appeal of Greenpeace's strategy to the general public by saying: "The average citizen thinks: 'Here I am dutifully recycling my garbage, and there comes big business and simply dumps its trash into the ocean.'"[14] In hindsight, Greenpeace's strategy has always been to try and to keep its message simple and connect it to the public's everyday experiences and values.

Despite the mounting protests and another attempt by Greenpeace to board the rig, Shell began towing the Brent Spar to its dumping site on June 11 as per schedule. During the following week the boycott of Shell gasoline stations in Germany was in full swing. Sales were off by 20 to 30 percent[15] and in some areas up to as much as 40 percent.[16] The mayor of Leipzig banned city vehicles from using Shell gasoline. Boycotts also spread to the Netherlands and Denmark. During the G-7 summit at Halifax, Canada, German Chancellor Helmut Kohl criticized Shell and the British government for persisting with the proposed deep-sea dumping. Two days later a firebomb exploded at a Shell gas station in Hamburg.

Shell had used high-powered water cannons to keep a Greenpeace helicopter from approaching the Brent Spar, but on June 16, two Greenpeace activists again succeeded in landing on the rig by arriving before the water cannons had been turned on. They managed to stay on Brent Spar while the rig was being towed to its chosen disposal site. On June 19, the German economics minister, Günter Rexrodt, announced that his ministry, too, would join the boycott. During this period, the German public received inconsistent messages from Shell. Although Shell Germany suggested that the project could be halted, Shell UK refused to stop the towing. Meanwhile in the United Kingdom, Prime Minister John Major was repeatedly attacked in Parliament but refused to reconsider the government's decision to approve Shell's proposal.

Shell blinks

After a meeting of the former Royal Dutch/Shell Group's managing directors in The Hague on June 20, Christopher Fay, chairman of the Shell UK, announced that Shell would abandon its plans to sink the Brent Spar. Fay stressed that he still believed that deep-sea disposal offered the best environmental option but admitted that Shell UK had reached an "untenable position" because of its failure to reach out and convince other governments around the North Sea.[17] Shell UK would now attempt to dismantle the platform on land and sought approval from the Norwegian authorities to anchor the Brent Spar temporarily in a fjord on the Norwegian coast. Environmentalists, not unexpectedly, received the decision with joy.

Aftermath

In the June 29 issue of *Nature,* two British geologists at the University of London argued that the environmental effects of Shell UK's decision to dump the Brent Spar in the deep sea would "probably be minimal."[18] Indeed, the metals of the Brent Spar might even be beneficial to the deep-sea environment. Disposing of the Brent Spar on land could pose greater risks to the environment.[19] Robert Sangeorge of the Switzerland-based Worldwide Fund for Nature said, "Deep-sea disposal seemed less harmful option."[20] He called the Brent Spar episode "a circus and sideshow that distracted from the big environmental issues affecting the world."[21] In response, however, a spokesperson for Shell UK reiterated that the company would stick to its decision to abandon deep-sea disposal.

The Brent Spar remained anchored in Erfjord, Norway. After an independent Norwegian inspection agency, Det Norske Veritas, had surveyed the contents of the Brent Spar, some doubts arose about Greenpeace's estimates of the oil sludge remaining on Brent Spar. Shell had previously estimated that the facility contained about 100 tons of sludge. Greenpeace had estimated 5,000 tons. One September 5, Greenpeace UK's executive director Lord Peter Melchett admitted that the estimates were inaccurate and apologized to Christopher Fay. Shell UK welcomed the apology and announced its intention to include Greenpeace among those to be consulted in its review of options and the development of a new BPEO.[22]

DOI: 10.1057/9781137388070.0014.

Brand lesson no. 4

Following the end of the Brent Spar episode, a number of public relations experts criticized Shell's handling of the protests and its decision to abandon its original plan. Mike Beard, former president of the Institute of Public Relations, commented, "Shell failed to communicate the benefits of the course they believed to be right; they lost what they believed to be their case; and now they're having to defend something they don't consider to be defensible."[23] In response, a production director of Shell, defended the company's decision not to involve environmental activist groups like Greenpeace: "Greenpeace does not have formal consultative status under the guidelines set out for an offshore installation proposal. Other stakeholders who represent a wide range of interests or who are accountable to their members are part of the process, and we consulted them."[24] Following the decision to halt the project, Shell started an advertising campaign admitting to its mistakes and promising change.

Although not commonplace for oil companies and environmental groups to work together, Shell might have been able to prevent damage to the brand by consulting with Greenpeace and other environmental organizations during its decision-making process. This action might also have prevented the boycott of Shell's retail stations. When confronted with a potential brand crisis, oil companies and all stakeholders, including environmental groups, should collaborate to address concerns, develop strategies, and implement plans for mutual benefit.

Notes

1. Andrew Critchlow. "Investment in North Sea Oil and Gas to Rise Sharply, Say Analysts." *The Telegraph* (online) (January 10, 2014).
2. Rudall Blanchard Associates Ltd. (for Shell UK Exploration and Production). *Brent Spar Abandonment BPEO*, (December 1994).
3. *Frankfurter Allgemeine Zeitung* (June 12, 1995).
4. Interview with Harald Zindler. *Greenpeace*, Germany (August 11, 1995).
5. Marlise Simons. "For Greenpeace Guerrillas, Environmentalism Is Again a Growth Industry." *New York Times* (July 8, 1995), p. 3.
6. Ibid.
7. Simon Reddy. "No Grounds for Dumping," *Greenpeace International* (April 1995).

8. *Interview* (August 11, 1995).
9. Cacilie Rohwedder and Peter Gumbel. "Shell Bows to German Greens' Muscle—Reversal of Plan to Sink Rig Shows Growing Clout of Environmentalists," *Wall Street Journal* (June 21, 1995), p. A15.
10. *Wall Street Journal* (July 7, 1995).
11. Bhushan Bahree and Kyle Pope. "Giant Outsmarted: How Greenpeace Sank Shell's Plan to Dump Big Oil Rig in Atlantic," *Wall Street Journal* (July 7, 1995), p. A1.
12. Ibid.
13. Accepted proposals of this conference were nonbinding.
14. David Baron. "Going Head to Head," *Stanford Social Innovation Review*, p. 38.
15. Bhushan Bahree and Kyle Pope. "Giant Outsmarted: How Greenpeace Sank Shell's Plan to Dump Big Oil Rig in Atlantic," *Wall Street Journal* (July 7, 1995), p. A1.
16. *Wirschaftswoche* (June 22, 1995).
17. *Financial Times* (June 21, 1995).
18. E.G. Nisbet and C.M.R. Fowler. "Is Metal Disposal Toxic to Deep Oceans?" *Nature* (June 29, 1995), pp. 375–715.
19. Ibid.
20. Bhushan Bahree. "Giant Outsmarted: How Greenpeace Sank Shell's Plan to Dump Big Oil Rig in Atlantic," *Wall Street Journal* (July 7, 1995), p. A1.
21. See footnote 11. *Wall Street Journal* (July 7, 1995).
22. *Financial Times* (September 5–9, 1995).
23. *Financial Times* (June 23, 1995).
24. Grant Jordan. *Shell, Greenpeace and the Brent Spar*. (New York: Palgrave Macmillan, 2001), p. 24.

9
The Tarnished BP Brand: From Texas City to Price Fixing

Abstract: *In addition to the Deepwater Horizon explosion and resulting oil spill in the Gulf of Mexico in April 2010, BP suffered additional blows to its new brand image. U.S. Government officials had fined the company over leaky pipelines in Alaska and attempts to control prices in the U.S. propane markets. All of these events seriously damaged BP's corporate reputation and brand image.*

Robinson, Mark L. *Marketing Big Oil: Brand Lessons from the World's Largest Companies*. New York: Palgrave Macmillan, 2014. DOI: 10.1057/9781137388070.0015.

Following BP's acquisition of Amoco and ARCO in the late 1990s, the company seized the opportunity to integrate all three brands into one. This was the beginning of the migration from British Petroleum to "beyond petroleum," which was unveiled in 2000 with much fanfare. This image makeover was intended to portray BP as an environmentally caring company, a company that would spend huge sums of money on solar and wind energy. But on March 23, 2005 at 1:20 p.m., BP's newly established brand suffered significant damage when part of the Texas City refinery exploded, resulting in a fire that killed 15 people and injured another 180. This incident alarmed the local community and resulted in financial losses exceeding US$1.5 billion. Additional disasters would soon follow: a second and third explosion and fire at the same refinery, allegations that BP illegally cornered part of the U.S. propane market, pipeline corrosion in Alaska, and media reports that some BP offices had been raided by European Commission officials over allegations of price manipulation. All of these events impacted the company's brand and corporate reputation despite the fact that some of the allegations were found to be untrue.

Texas City refinery explosion

At the time of the Texas City explosion, the refinery was BP's largest and most complex refinery with a rated capacity of 460,000 barrels per day and an ability to produce 11 million gallons of gasoline a day. Following the incident BP attempted to use the "shift the blame" image restoration strategy by stating that the refinery was owned and operated by BP Products North America and not by its London-based parent. As readers will recall from the Exxon *Valdez* incident described earlier, the shifting the blame strategy is an attempt by the company to shift the blame from itself onto other stakeholders including contractors as well as federal and state governments who are in charge of oversight. When BP tried to shift the blame to one of its own subsidiaries—BP Products North America—it failed miserably. Ultimately, the London-based parent *did* own the North American unit. Four months later, on July 28, a second explosion and fire occurred at Texas City; this time there were no injuries or fatalities. On August 10, a third incident occurred. In that incident, BP reported a leak in the gas oil hydro-treating unit and ordered local residents adjacent to the facility to seek shelter as a precaution.

In the weeks and months following all three incidents, the U.S. Chemicals Safety and Hazard Investigation Board (CSB) launched several investigations. These investigations uncovered numerous safety and personnel issues, which in turn, led to a fine of US$21.3 million from the U.S. Occupational Safety and Health Administration (OSHA). OSHA conducted its own investigation and uncovered 303 willful violations.[1]

In hindsight, the CSB report was the most damning to the BP brand. While describing the many technical details of the March 25 explosion and fire, the CSB identified management and operational failures for the catastrophic incident. Investigators had found that BP was aware of problems at the facility but implemented reforms aimed at improving compliance with procedures and reducing occupational injuries while ignoring potentially large safety risks. According to CSB Chairwoman Carolyn Merritt's statement at the time, "Unsafe and antiquated equipment designs were left in place, and unacceptable deficiencies in preventive maintenance were tolerated."[2] Chairwoman Merritt went on to say, "BP implemented a 25 percent cut on fixed costs from 1998 to 2000 that adversely impacted maintenance expenditures and infrastructure at the refinery."[3] The CSB report also found the refinery's central training staff was reduced from 30 employees in 1997 to 8 in 2004, while the training department budget was cut in half from 1998 to 2004.[4]

On a more positive note, BP's management response to the March 25 incident was initially swift but in the long term, not sustainable. Lord John Browne, then BP's CEO immediately flew to the refinery site to take responsibility and to set up a victim's compensation fund. But not all of BP's executives felt the same level of urgency, as did Lord Browne. A colleague had emailed John Manzoni, then BP's head of marketing and refining, suggesting that he should travel to the refinery to assist. Manzoni's reply was terse. He appeared to be also perturbed when Lord Browne ordered him to interrupt his vacation to go to Texas City: "I arrived in Texas City at 3:00 a.m. along with Lord Browne and we spent the entire day there—at the cost of a precious day of my leave."[5]

Cornering the U.S. propane market

BP's problems were not limited to safety and occupational issues stemming from the Texas City incident. During 2004, there were allegations that the company illegally manipulated propane prices by attempting to corner the market for the gas. The propane case was one of a series of enforcement actions to impact BP's brand which involved its trading operations in the United States. The Commodity Futures Trading Commission (CFTC) alleged that the company had cornered the market in TET propane—the type commonly used to heat homes—by purchasing a "dominant and controlling long position." This action forced buyers to come to BP for supply at prices dictated by the company rather than the free and open market. Three years after the allegations were made, BP settled the charges for US$303 million, naturally admitting no wrongdoing.

Pipeline corrosion in Alaska

Oil companies spend millions of dollars building pipelines to transport crude oil from where it is produced to refineries where it is refined into retail products such as gasoline/petrol and jet fuel. Large sums of money are also spent on maintaining these pipelines based on legal mandates from governments. One of the major issues in pipeline maintenance is corrosion.

Alaska's Prudhoe Bay field is the site of one of the largest oil discoveries made in North America during the 1970s. Since then, the field has been one of the largest producing oil fields in the world. In early August 2006, the discovery of unexpected levels of pipeline corrosion and a spill forced the partial shutdown of the production field. At the time this incident occurred, BP was still struggling to complete corrosion checks mandated by the U.S. government after a 200,000 gallon crude oil spill that previous March.

As would be expected, a U.S. Congressional hearing took place soon after the August 2006 discovery. On September 7, 2006, executives from BP spent the day testifying before a U.S. Congressional subcommittee. The oversight committee, the House Energy and Commerce Committee, was there to investigate the BP's failed corrosion maintenance program for the Prudhoe Bay pipelines.

At the outset of the hearings, House members attempted to question BP's former manager in charge of corrosion prevention, Richard Woollam. Woollam, who was on paid leave at the time, refused to answer the committee's questions by invoking his Fifth Amendment right against self-incrimination. Over one year earlier, in January 2005, Woollam was removed from his position as manager of the corrosion prevention staff at Prudhoe Bay after complaints surfaced that his aggressive management style deterred employees from bringing corrosion issues to his attention. The committee was also interested in finding out more about rumored budget and staff reductions to the corrosion program. In addition, there were rumors regarding an Alaskan government report that was diluted after the company complained that its anticorrosion program had been unfairly criticized.

As part of the questioning process, the subcommittee focused on BP's failure to use what are called maintenance pigs and diagnostic pigs on some Prudhoe Bay pipelines for the 14 years previous to the corrosion discovery in 2006. A pig is a mechanical device that travels inside the pipeline from beginning to end, checking for microscopic evidence of rust and corrosion. In contrast, the subcommittee members heard testimony that the much larger trans-Alaskan pipeline, also built during the early 1970s, is being inspected with a diagnostic pig every three years to detect for corrosion and a maintenance pig is run through the pipeline at least every 14 days to clean it. In response, BP's assertion that the failure to pig the lines was not due to cost, but was because the company didn't think the lines were vulnerable to corrosion. Steve Marshall, then president of BP's Alaska subsidiary, took the lead in answering the oversight committee's questions. Marshall defended BP's corrosion program, but did admit that a mentality of cost reduction fostered by record low oil prices in the late 1990s may have contributed to a culture of lax maintenance.

During the subcommittee's hearings, internal BP documents were presented which indicated the position of senior corrosion engineer for BP's Alaska operations had been vacant for at least 18 months. In an April 2005 report to BP's Alaskan management team, John Baxter, then director for engineering at the company, warned BP's management team that both Woollam's and the position of senior corrosion engineer, vacant since December 2004, needed to be filled as urgently as possible so as not to reduce the effectiveness of the corrosion prevention team. Following the Congressional hearings, BP announced that it had retained the services of three outside corrosion experts to evaluate the corrosion control program in Alaska and to make recommendations for improvements.

The European Commission office raid

In any organization, allegations over price manipulation made by external stakeholders should be taken seriously. In May 2013, BP, Royal Dutch Shell, and Norway's Statoil's offices were raided by European Commission (EC) officials over allegations the three major oil companies manipulated prices of oil so as to profit from retail prices. The commissions' statement that the alleged manipulation could have a large impact on the reported price of crude oil and refined products could also harm consumers. If in fact, the companies were to have been found to fix prices, the EC can levy fines of up to 10 percent of a company's global sales/turnover.

According to Fitch, the corporate equities rating service, major oil companies sometimes carry between US$10 billion to US$20 billion of cash on their balance sheets, so any fines levied would be easily manageable.[6] However, Fitch was also quoted as stating: "The larger problem is the longer term risks from damage to their reputation" if any of the three companies were found to have distorted prices.[7]

An oil company's trading business typically sells not only oil and other refined products, but also bio-fuels, chemical feed-stocks, and liquefied natural gas (LNG) products. The trading business can oftentimes be profitable but highly volatile. In one example, BP's 2013 first quarter financial results exceeded the market's expectations largely due to a strong contribution from its trading arm. Financial results from an oil company's trading division are often highly secretive as these results are not required by law to be separated from typical upstream (exploration and production) and downstream (retail) financial data.

Brand lesson no. 5

When BP set out to change its corporate brand in 2000, it was seen by many as an effort to not only incorporate the brands of Amoco, ARCO, and BP into one brand, but also to change its perception from an oil company to one that was more in tune with environmental concerns. At the outset of its rebranding strategy, the company broke ranks with other oil companies by becoming a member of numerous environmentally focused organizations and actively promoting its belief that climate change was occurring due to the burning of fossil fuels. BP also made investments in solar and wind energy as well as bio-fuels to demonstrate its commitment. And for a short time, the strategy seemed to resonate with stakeholders.

When a company possesses strong brand equity, internal and external events will likely do little damage to the company's brand and corporate reputation. In BP's case, the company's rebranding effort was initially successful but there was no strong brand equity from which the company could draw from to recover from the multiple crises.

Before beginning a rebranding initiative, oil companies should first conduct a brand audit to gauge the strength of the company's brand vis-à-vis its competitors. A brand audit will reveal the company's brand strengths and weaknesses. Building a strong brand now should be one of the most important initiatives an oil company should undertake.

Notes

1. U.S. Chemical Safety and Hazard Investigation Board. "*Investigation Report: Refinery Explosion and Fire.*" (March 23, 2005), www.csb.gov/assets/1/19/CSBFinalReportBP.pdf.
2. Ibid.
3. Ibid.
4. Ibid.
5. Sheila McNulty. "Manzoni Holiday Hit by BP Explosion," *Financial Times* (October 13, 2006), FT.com.
6. Guy Chazan. "Oil Companies Face Public and Political Backlash if Found Guilty," *Financial Times* (May 15, 2013), FT.com.
7. Ibid.

10
Chevron versus Ecuador: How a Strong Brand Defends Itself

Abstract: *Following 20 years of oil drilling operations in Ecuador, native tribes in the Amazon Rain Forest sued Texaco's Texpet subsidiary over environmental pollution. The company undertook its own cleanup operations before exiting the country in 1992. As a result of Chevron's acquisition of Texaco, the same native tribes sued the newly formed ChevronTexaco for $19 billion. Chevron has pursued a Standard Oil-style defense refusing to financially settle the lawsuit.*

Robinson, Mark L. *Marketing Big Oil: Brand Lessons from the World's Largest Companies.* New York: Palgrave Macmillan, 2014. DOI: 10.1057/9781137388070.0016.

In their unending quest to locate new oil and natural gas fields, Big Oil companies often find themselves working in physically inhospitable and politically and economically risky geographical areas. These hotspots range from the Delta region in Nigeria, the cold climate of eastern Russia, the oil sands regions of Canada, and anti-U.S. countries in Latin America such as Venezuela and Ecuador.

As part of this search for oil and gas, companies must also be willing to acquire competitors that own promising oil fields they themselves do not own but desire. In purchasing competitors, acquiring companies typically conduct a due diligence review regarding various aspects of the target company such as financial statements, geographical operations, oil and gas reserves, but oftentimes overlook the negatives including potential environmental liabilities. When one company acquires another, the acquiring company takes on any liabilities of the acquired company. A case in point was Chevron's acquisition of its former competitor Texaco in 2001. This acquisition led to Chevron's inheritance of Texaco's legal case involving allegations of dumping toxic oil drilling and production waste into Ecuador's Amazon basin. The impact the case has had on Chevron's brand will be described shortly, but first a little history is necessary to begin the story.

Texaco's operations in Ecuador

In 1964, the Latin American nation of Ecuador granted an exploration concession to a consortium of U.S. oil companies including a Texaco subsidiary (Texaco Petroleum or Texpet) to prospect for oil in the country's Amazonian forest, which it discovered three years later.[1] By 1974, Ecuador's government-owned oil company PetroEcuador became the majority owner in the consortium with Texpet, and gradually increased its ownership stake until it assumed full operational control in 1990. In 1992, the consortium agreement expired and Texaco ended its Ecuadorean operations. In 1993, Ecuadorean residents near the dumping sites filed a class-action lawsuit in New York alleging that the company polluted the local habitat with oil waste. A second related lawsuit was filed during late December 1994 by Peruvian Indians who claimed Texaco's dumping of toxic drilling waste in Ecuador found its way into a river that flowed into Peru. That lawsuit sought damages in the amount of US$1 billion as well as class action status.[2]

In 1998, the Ecuadorean government formally "absolved, liberated, and forever freed" Texaco from "any claim or litigation by the Government of Ecuador concerning the obligations acquired by Texpet."[3] An independent environmental audit conducted shortly thereafter concluded that Texpet had abided by existing environmental regulations and industry practices, though it recommended that Texaco invest US$13.2 million to remediate localized damages. In reality, Texaco ended up spending US$40 million on the project to clean up 161 abandoned drilling pits. From this point on, and according to the release signed by the government, PetroEcuador was legally required to clean up the rest of the pits, but didn't do so.[4]

During October 2003, two years after Chevron acquired Texaco, the lawsuit looked favorable to the Ecuadorean plaintiffs when U.S. courts ruled that Ecuador has primary jurisdiction in the case. The plaintiff's attorneys were quick to point out that this was the first time any U.S. court agreed to recognize as binding an international court's authority in deciding whether a U.S. company has damaged the environment.

Chevron goes on the offensive

Upon entering the legal proceedings, Chevron—formerly Standard Oil of California—took a page from the Standard Oil/John D. Rockefeller, Sr. playbook by stating: "We're not paying and we're going to fight this for years if not decades into the future."[5] Readers will recall this was the identical strategy used by Exxon following the *Valdez* spill in 1989; plaintiffs file a suit against a company for environmental damage while the company appeals any court ruling made against it. This legal maneuvering takes several decades if not more to complete until the company is able to whittle down any settlement payment to the lowest legally possible.

During the lengthy legal process, an independent expert appointed by an Ecuadorean court recommended the judge award the plaintiffs US$27 billion in damages.[6] At the time, this amount would have represented roughly one-tenth of the company's 2008 revenue.[7] This amount was later revised upward to US$113 billion.[8]

One of Chevron's offensive strategies brought to the Ecuador litigation was to sue Ecuador's government under international trade law. The company had requested arbitration in September 2009, claiming the Ecuadorean government was interfering in the long-running lawsuit brought by local indigenous groups over Texaco's alleged damage to the local environment. Ecuador had earlier filed a lawsuit against Chevron in the U.S. District Court in Manhattan in December 2009, seeking to stop an arbitration by Chevron under the rules of the United Nations Commission on International Trade Law.

In an effort to protect it from what the company believed was an illegal stacked deck, Chevron counter sued under the terms of a 1997 trade pact between the U.S. and Ecuador. The suit amounted to a request for arbitration through a process set up by the United Nations Commission on International Trade Law to adjudicate disagreements. This arbitration process was separate from the original lawsuit and under the international trade pact with the United States the Ecuadorean government must accept the arbitrator's rulings as binding under international law. The company further implemented a website to rebut plaintiff's claims in an effort to revoke Ecuador's trade privileges.

After months of reviewing Chevron's request for arbitration, a U.S. District Court Judge ruled on March 11, 2010, that Chevron's claim that Ecuador's government had infringed on its right to due process is a claim that can be arbitrated. Following the court's decision, the international arbitration panel ruled that Ecuador must pay Chevron up to US$700 million in damages and interest because its courts took too long to rule on lawsuits brought by the company.[9]

The international arbitration panel that was reviewing Chevron's suit against Ecuador reached a decision on September 18, 2013. In a huge victory for the company, the international arbitration tribunal issued a partial ruling in favor of Chevron and found the oil company wasn't liable for collective environmental damage claim in Ecuador.[10] In stating its decision-making process, the tribunal found Chevron and Texaco are "releasees" under agreements reached in 1995 and 1998 and that Chevron can enforce its rights as a releasee.[11] In an additional decision, the tribunal also ruled that the Ecuadorean government had settled all public interest or collective environmental claims, including collective claims asserted by third parties.[12]

Chevron scored an earlier public relations victory when the U.S. CBS weekly news program *60 Minutes* aired its story titled "Amazon Crude" on May 3, 2009. During the top-rated news show, Chevron vigorously defended itself against allegations that it had dumped toxic waste into the Amazon rainforest. As part of its defense against the television broadcast that the company deemed to be a smear campaign, Chevron sought advice from the *Columbia Journalism Review* (CJR), which assigns a reporter to investigate complaints from businesses about unfair treatment by the media. The CJR published a detailed analysis by Martha M. Hamilton, part of which says: "Amazon Crude is an interesting study in appearance and reality in a TV news documentary and how the one doesn't always add up to the other."[13] In her analysis, Hamilton went on to describe a scene from "Amazon Crude" in which a man scoops up a thick, black substance—presumably crude oil—and states that he is unable to drink water from his well. The implication is clear: the oily goo was toxic waste from one of Texaco's drilling and production sites. As it turned out, the thick substance is not oil and the well wasn't Texaco's responsibility. Even more revealing, the Ecuadorean man wasn't even a plaintiff in the case. State-owned PetroEcuador's complete ownership of the sites at the time was not revealed during the *60 Minutes* story.

Legal tug-of-wars continued post the *60 Minutes* news episode when in early March 2011 a federal judge in Manhattan issued a preliminary injunction blocking the plaintiffs and their lawyers from attempting to enforce an Ecuadorean court's ruling that they cannot seize the oil giant's assets anywhere outside of Ecuador.[14] Judge Lewis A. Kaplan wrote that Chevron had presented ample and credible evidence that the Ecuador trial had been tainted by fraud and corruption. Kaplan went on to say that Chevron faced "irreparable harm" if the plaintiffs were allowed to pursue their strategy.

Chevron back on the defensive

Chevron's legal wins since the *60 Minutes* news report were increasing. It had successfully defended itself against an attack by a major U.S. news program, brought suit against a small Latin American nation via an international arbitration panel, and brought attention to the corruption and fraud against Ecuadorean judges and the plaintiffs' legal team and environmental consultants. However important these victories were to Chevron, helped in part by Judge Kaplan's earlier ruling in Manhattan, they were short lived. In early January 2012, a U.S. appeals court discarded a bid by Chevron to block a group of Ecuadoreans from collecting a multi-billion judgment over environmental damage. According to U.S. Circuit Judge Gerard E. Lynch's 30-page order, there was "no legal basis" for the injunction until the plaintiffs seek to enforce the judgment in a court "governed by New York or similar law."[15] But by October 2012, the Second U.S. Court of Appeals in New York ruled that Judge Kaplan couldn't issue such a global order. Although Chevron appealed the Second Circuit ruling, the Supreme Court declined to hear the case.[16]

Plaintiffs seek damages in Canada

As the plaintiffs' legal team began to realize that Chevron was unlikely to make any settlement payments in Ecuador, the teams' lawyers devised a new strategy: seeking damages in other countries where the company owned sizeable assets. The most viable locations for this asset squeeze were Argentina and Canada. Through legal action, the plaintiffs were able to freeze some of the company's assets in Argentina but also sought legal action in Canada during May 2013, particularly in Ontario.

Although the plaintiffs achieved some success with this strategy in Argentina, Chevron prevailed in Ontario, when Judge D.M. Brown of the Superior Court of Justice ruled that Chevron's subsidiaries are legally separate from the company and not subject to the Ecuador court's verdict. According to the judge's ruling, Chevron owned no assets within the court's jurisdiction, adding: "The evidence disclosed that there is nothing in Ontario to fight over."[17] Seven months later, in December 2013, the Canadian Court of Appeals reversed the lower court's ruling stating, "The Ecuadorean plaintiffs should have an opportunity to attempt to enforce the Ecuadorean judgment in a court where Chevron will have to respond to the merits."[18]

Plaintiffs lawyers withdraw

In May 2013, a law firm representing the Ecuadorean plaintiffs reluctantly withdrew from the case stating it was owed US$1.8 million in legal fees and that its clients were unable to pay. A second law firm also withdrew from the case. This second law firm representing Steven Donziger, a legal expert for the Ecuadorean plaintiffs who was accused of fraud by Chevron, said Donziger owes it US$1.4 million and also asked to withdraw.[19] Commenting on their withdrawal from the case, Donziger's lawyers, San Francisco-based Keker & Van Nest LLP, said they "should not be made a slave to this impossible situation," which they also called a "Dickensian farce."[20] The resignation of these prominent law firms demonstrates how Chevron's aggressive legal strategy impacted the plaintiff's legal team's ability to stay on the course.

Ecuador's last gasp for breath

Following the Ecuadorean government and plaintiff's significant losses against Chevron, Ecuador tried one last legal tactic. In mid-November 2013, Ecuador's highest court—the National Court of Justice (CNJ)—cut the fine a lower court imposed on Chevron from US$19 billion to a still mammoth US$9.5 billion.[21] The CNJ dismissed the most blatant part of the lower-court ruling, in which the judge doubled the damages because Chevron failed to apologize. The court said that the judge had no legal grounds to impose this kind of condition.[22] U.S. District Judge Lewis A. Kaplan issued a more recent legal ruling on March 4, 2014. In the 485-page decision, Judge Kaplan wrote: "The saga of the Lago Agrio case is sad."[23] Kaplan went on to write that the plaintiffs, indigenous people living in Ecuador's Amazon, "received the zealous representation they wanted, but it is sad that it was not always characterized by honor and honesty as well."[24]

Brand lesson no. 6

One of the characteristics of a strong brand is the ability to aggressively defend itself against any and all attackers. In Chevron's case, it has the legal muscle and financial resources to outlast any plaintiff's group. Big Oil stands alone at the top of all industries in being able to fight for what it believes is right. John D. Rockefeller, Sr. couldn't have been more proud.

Notes

1 Bret Stephens. "Amazonian Swindle," *Wall Street Journal* (October 30, 2007), p. A18.
2 "Peruvian Indians Charge Texaco with Oil Dumping," *Wall Street Journal* (December 29, 1994), p. A6.

3 Ibid.
4 "The Americas: Justice Or Extortion? Ecuador, Chevron, and Pollution," *The Economist* (May 23, 2009), p. 42.
5 Ben Casselman. "Chevron Expects to Fight Ecuador Lawsuit in U.S.: As Largest Environmental Judgment on Record Looms, the Oil Company Reassures Shareholders It Won't Pay," *Wall Street Journal* (July 20, 2009), p. B3.
6 Ibid.
7 Ibid.
8 Eric Watkins. "Chevron: No Scientific Basis for $113 Billion Ecuador Claim," *Oil & Gas Journal* (January 3, 2011), p. 31.
9 Ben Casselman. "Ecuador to Pay Chevron Damages," *Wall Street Journal* (March 31, 2010).
10 Nathalie Tadena. "Tribunal Finds Chevron Not Liable for Collective Environmental Claims," *Wall Street Journal Online* (September 18, 2013).
11 Ibid.
12 Ibid.
13 Bob Tippee. "Chevron Elicits Rebuke of *60 Minutes*' Smear," *Oil & Gas Journal* (April 26, 2010), p. 64.
14 Ben Casselman and Chad Bray. "Chevron is Granted Ecuador Injunction," *Wall Street Journal* (March 8, 2011) p. B1.
15 Bray. "U.S. Court Rejects Chevron Bid to Block Ecuadorean Judgment Collection," *Wall Street Journal Online* (January 26, 2012).
16 Barrett. "Chevron Fails to Squelch US$19 Billion Ecuador Verdict," *Bloomberg Businessweek (businessweek.com)* (October 9, 2012).
17 Gilbert and Angel Gonzalez. "Court Deals Blow to Ecuador Plaintiffs in Chevron Case," *Wall Street Journal (Online)* (May 2, 2013).
18 "Ecuadorean Plaintiffs Can Attempt to Seize Chevron's Canada Assets," *Wall Street Journal Online* (December 17, 2013).
19 Gilbert. "Plaintiffs' Lawyers in Chevron Case Seek to Withdraw," *Wall Street Journal Online* (May 4, 2013).
20 Ibid.
21 "The Last Word, With More to Follow: Oil in Ecuador," *The Economist (Online)* (November 15, 2013).
22 Ibid.
23 Steven Mufson. "U.S. Judge Rules for Chevron in Ecuador Case," *Washington Post* (March 5, 2014). http://www.washingtonpost.com/business/economy/us-judge-rules-for-chevron-in-ecuador-pollution-case/2014/03/04/ec828d00-a3bb-11e3-84d4-e59b1709222c_story.html.
24 Ibid.

11
A "Shell" Game for Investors

Abstract: *On January 9, 2004, a minor news story appeared in the financial media regarding Royal Dutch Shell's announcement regarding its oil reserves reporting. Within three months of this seemingly innocuous announcement the after tax income of the corporation for the preceding four years had been recalculated and reduced by almost US$450 million, roughly US$100 million per year for each of the four years in question. The incident created a negative brand image for Royal Dutch Shell.*

Robinson, Mark L. *Marketing Big Oil: Brand Lessons from the World's Largest Companies.* New York: Palgrave Macmillan, 2014. DOI: 10.1057/9781137388070.0017.

80 Marketing Big Oil

Every year since the 1970s, publicly-traded oil and gas companies in the United States have been required by the Securities and Exchange Commission (SEC) to disclose details pertaining to their production activities, estimates of their proved oil and gas reserves, and estimates of the present value of the future cash flow streams those reserves are expected to generate. When accounting regulators in the U.S. set the current disclosure rules in 1982, they decided that these figures—estimates—should only be reported as supplementary information outside companies' official financial statements. The reason was that the estimates weren't reliable enough to justify the cost of having them independently audited. Over the years, reserves estimates have become commonplace in oil company financial statements and are closely followed by investors as a measure of future production and profits. To understand reserves accounting, a brief overview is provided below.

Proven oil reserves are those reserves claimed to have a *reasonable certainty*, normally of at least a 90 percent level of confidence, of being recoverable under existing economic and political conditions, and by utilizing existing technology. Proven reserves are also known in the industry as 1P. Proven reserves are further subdivided into "proven developed" (PD) and "proven undeveloped" (PUD). Until December 2009 1P proven reserves were the only type the U.S. Securities and Exchange Commission allowed oil companies to report to investors. Oil companies listed on U.S. stock exchanges must substantiate their claims, but many governments and national oil companies (NOCs) do not disclose this verifying data to support their claims. Since January 2010 the SEC has allowed companies to further provide additional and optional information declaring 2P (both proven and probable) and 3P (proven + probable + possible) provided the evaluation is verified by qualified third party consultants, though many companies choose to use 2P and 3P estimates only for internal purposes.

The calm before the storm

On Friday January 9, 2004, a minor news story appeared in the financial media regarding Royal Dutch Shell's announcement involving its reserves reporting. The announcement ended with a statement that a teleconference would be held on the same day which would be hosted by several executives of the Royal Dutch/Shell Group of Companies ('Shell')—

Simon Henry, Head of Group Investor Relations; Mary Jo Jacobi, Vice President of Group External Affairs; and John Darley, Exploration and Production Technical Director. The announcement included the following phrase, "... some proved hydrocarbon reserves will be recategorized." The company's statement sought to reassure its investors by further adding that there would be no material impact on financial statements.

Just under a month later, the company announced its yearend financial results for 2003. Highlighted by the Chairman were the following aspects of performance:[1]

- Reported net income of US$12.7 billion in 2003 was 35 percent higher than in 2002
- The Group's earnings on an estimated current cost of supplies basis for the full year were a record of US$13.0 billion (46 percent higher than last year)
- Final dividends proposed of €1.02 per share for Royal Dutch and of 9.65p per share for Shell Transport, increasing above inflation. The increase in US dollar terms, at current exchange rates, exceeds ten percent and over the past three years has risen by 28 percent.

Within three months of this seemingly innocuous announcement, the after tax income of the corporation for the preceding four years had been recalculated and reduced by almost US$450 million, roughly US$100 million per year for each of the four years in question.[2] How could this financial brand crisis event have occurred at such a well-managed and financially disciplined company?

The eye of the storm

Following the announcements of the reserves reductions, it was revealed that some company executives were aware of the growing evidence that the oil giant had greatly over estimated its oil and natural gas reserves. Sir Philip Watts, the company's chairman, and Walter van de Vijver, Shell's chief of exploration and production, acted to keep the information secret according to documents revealed during the company's internal investigation. In an email dated November 9, 2003, van de Vijver wrote to his boss, Sir Philip, saying that "I am becoming sick and tired of lying about the extent of our reserves issues and the downward revisions that need to be done because of far too aggressive/optimistic bookings."[3]

According to the report, van de Vijver was seen as a first-rate executive who wanted to reveal the controversy to investors, repeatedly complaining about the issue internally to his boss. Yet the report showed that he ultimately failed and that he eventually sided with Sir Philip. Additionally, he neglected to take his formal complaints outside the executive suite. In one additional note to Shell's senior executives, van de Vijver warned of the trouble executives were facing in trying to keep the market "fooled," according to the report.

Some of the documents revealed during the company's internal investigation include briefing material circulated to the company's top management team, emails, and private notes between the two executives. The correspondence reveals that senior executives were aware of serious reserves-accounting problems for several years but didn't reveal the information to regulators, the public, or their own boards (remember at the time, Shell was a combination of two companies, each with a separate board, Royal Dutch Petroleum and Shell Transport and Trading). Furthermore, the documents also pointed to a growing discrepancy between what company executives knew and what Shell was disclosing to the investing public.

In February 2002, van de Vijver forwarded a memo to the executive committee detailing potential reserves overbookings of some one billion barrels of oil equivalent (boe) (meaning a combination of oil, natural gas, and other liquids acquired through drilling operations). In addition, the memo stated that another 1.4 billion boe were at risk. According to the report, a second presentation in July 2002 made to the executive committee failed to address Shell's non-compliance with SEC rules regarding reserves bookings. The July presentation comes across as an attempt to "play for time" in hopes that some internal or external development would "justify or mitigate the existing reserve exposure."[4]

In a note to the executive committee on September 2, 2002—with a copy sent to Boynton— van de Vijver described "dilemmas" facing Shell's exploration and production (E&P) division: "Given the external visibility of our issues...the market can only be "fooled" if (1) the credibility of the company is high, (2) medium and long-term portfolio refreshment is real and/or (3) positive trends can be shown on key indicators."[5] The memo goes on to say: "We are struggling on all three criteria."[6] Van de Vijver later attempted to excuse his own conduct by shifting the blame to Shell's culture and targeted Judy Boynton, the chief finance officer, for ultimately preventing his efforts to disclose the problems inside Shell.

The impact

According to the *Cautionary Note to Investors* that Shell included at the time in its financial documents, which were also available in its website: "The United States Securities and Exchange Commission permits oil and gas companies, in their filings with the SEC, to disclose only proved reserves that a company has demonstrated by actual production or conclusive formation tests to be economically and legally producible under existing economic and operating conditions."[7]

It is apparent that this requirement was completely ignored and that the company carelessly overstated its oil reserves over an extended period of time, from 1997 to 2002. According to the SEC, the company overstated proved reserves reported in its 2002 Form 20-F filing by 4.47 billion barrels of oil equivalent or approximately 23 percent. The company also overstated the standardized measure of future cash flows reported in this filing by US$6.6 billion. In its defense, Shell corrected these overstatements in an amended filing on July 2, 2004, which reflected the degree of the overstatements for the years 1997–2002. It is important to point out that modifying billions of barrels of oil equivalent in percentages ranging from 16 to 25 percent is a significant readjustment as seen in Table 11.1.

These proved reserves overstatements resulted in the company overstating the standardized measure of future cash flows by US$27.3 billion.[8]

TABLE 11.1 *Shell's reserves overstatements*

Year	Proved reserves overstatement (boe)	(%) Overstatement	Standardized measure overstatement (US$)	(%) Overstatement
1997	3.13	16	N/A	N/A
1998	3.78	18	N/A	N/A
1999	4.58	23	7.0 billion	11
2000	4.84	25	7.2 billion	10
2001	4.53	24	6.5 billion	13
2002	4.47	23	6.6 billion	9

Note: Royal Dutch Shell. Securities and Exchange Commission Form 20-F for years 1997–2002.
Source: Shell Form 20-F 1998, 1999, 2000, 2001.

During this same period it was also found that Shell materially misstated its reserves replacement ratio (RRR), a key performance indicator in the oil and gas industry. A common definition of RRR is "a metric used by investors to judge the operating performance of an oil and gas exploration and production company. The reserve-replacement-ratio measures the amount of proved reserves added to a company's reserve base during the year relative to the amount of oil and gas produced. During stable demand conditions a company's RRR must be at least 100 percent for the company to stay in business long-term; otherwise it will eventually run out of oil."[9] If Shell had properly reported proved reserves, its RRR would have been 80 percent rather than the 100 percent for the five-year period.

Management misbehavior

As the U.S. SEC's inquiry outlined, Shell's overstatements of proved reserves as well as its delay in correcting the overstatement resulted from its desire to create and maintain the appearance of a strong RRR, i.e., a positive image to investors and the general public. Although Shell was warned on several occasions prior to the fall of 2003, the company rejected the warnings as immaterial or unduly pessimistic. According to one publication, there was an ongoing culture of arrogance among the senior managers of the company, which made this kind of contempt for regulation unsurprising.[10]

Bearing in mind the previously described analysis of Shell's actions, it is worth reviewing the company's key financial data, share price, and dividends paid because of its impact on year-end performance compensation. Over the 1998–2002 period Shell's Return on Equity (ROE) ratio increased from 0.6 to 16.2 which indicates it went up by an astonishing 2600 percent![11] Furthermore, its earnings per share (Net Income per Share) of Royal Dutch valued in euros per ordinary share was similar to the increase of the ROE ratio of 2500 percent.[12] As is the case in many publicly traded companies, these financial measurements form the basis of year-end bonuses and are therefore key motivators for managers to utilize aggressive techniques—such as overestimation of reserves—to support these earnings. Such strategies have an unusually short shelf life as the market eventually uncovers the distorted numbers and the schemes used to create them. For years 1998 through 2001, Royal Dutch

TABLE 11.2 *Shell's number of shares under option at the end of the year granted to executives and other employees*

Year	Royal Dutch	Shell Transport and Trading
1998	7,078,450	40,741,224
1999	6,063,190	34,434,024
2000	10,214,460	41,936,904
2001	20,401,000	65,012,000

Note: Royal Dutch Shell. Securities and Exchange Commission Form 20-F for years 1997–2002.
Source: Shell Form 20-F 1998, 1999, 2000, 2001.

Shell's number of shares under option at the end of the year granted to executives and other employees are shown in Table 11.2.

It is typical in today's business arena that publicly traded companies increase the number of share options given each year to management. In Shell's case, this number increased by 188 percent from 1998 to 2001 in the case of Royal Dutch and 60 percent in the case of Shell Transport and Trading. If there is a relationship between issuing misleading oil reserves information and company image it would appear to be in management's interest to continue to issue such information towards maintaining a high share price by any means available—unethical or otherwise.

Brand lesson no. 7

A current trend in big business is that bad news—although often known in its completion by the company's executives—is only released into the public domain on a piecemeal basis, as if this makes the news more palatable.[13] Consequently, investors are never quite sure if all of the relevant information has been issued, creating a brand image that is not as financially positive as it could be.

This episode served as a catalyst for several management changes that sought to reassure investors and other company stakeholders. The most significant casualty involved the Chairman, Sir Philip, who later resigned from his position. Sir Philip was not the only victim of the reserves scandal, as Walter van de Vijver, Shell's chief of exploration and production was later ousted from his position, while in April 2004, Shell's Chief Financial Officer, Judy Boynton, stepped down from her post. The story does not end there as Shell experienced several other downward reserves adjustments. During March 2004, Shell announced it was reducing its oil and gas reserves once again although by a far smaller margin than in January. All told, Shell reclassified its reserves estimates five times in a little over a year.[14] The reduction in reserves and in financial adjustments seriously damaged Shell's image, reputation, and financial credibility.

Notes

1. *Shell Press Release* (February 5, 2004).
2. See the report to the *Group Audit Committee and Reserves Recalculation Review* (April 19, 2004).
3. Chip Cummins and Alexei Barrionuevo. "Former Shell Officials Clashed as They Hid Reserves Problems," *Wall Street Journal* (April 20, 2004), p. A1.
4. Ibid.
5. Ibid.
6. Ibid.
7. *Shell Press Release* (February 5, 2004).
8. Royal Dutch Shell. Securities and Exchange Commission Form 20-F for years 1997–2002.
9. www.investopedia.com, accessed July 13, 2013.
10. I. Cummins and J. Beasant. *Shell Shocked.* (Mainstream Publishing: London, 2005).
11. David Crowther and Esther Ortiz Martinez. "No Principals, No Principles and Nothing in Reserve: Shell and the Failure of Agency Theory," *Social Responsibility Journal* 3(4) (2007), pp. 4–14.
12. Ibid.
13. Royal Dutch Shell. Securities and Exchange Commission Form 20-F for years 1997–2002.
14. Paula Dittrick. "Reserves Reporting Debate Stirs Movement for Reform," *Oil & Gas Journal* (June 20, 2005), p. 20.

Part III
Marketing Strategies and Brand Building

12
Marketing and Advertising Innovation at Mobil Oil

Abstract: *In an attempt to break out of the often vicious and unprofitable price wars that oftentimes erupt from local retail gasoline competition, Mobil Oil launched several innovative marketing and advertising programs. The first involved the "advertorial" or advocacy editorial initiative that began in the early 1970s. The second was an in-depth market research project during the mid-1990s involving interviews with 2,000 customers to gain new insights into the types of services gasoline customers wanted. Both of these initiatives enabled Mobil Oil to create a new brand image and to increase profits by five percent.*

Robinson, Mark L. *Marketing Big Oil: Brand Lessons from the World's Largest Companies.* New York: Palgrave Macmillan, 2014. DOI: 10.1057/9781137388070.0019.

Two of the opportunities Big Oil companies were able to capitalize on following the dissolution of Standard Oil in 1911 involved the growing popularity of gasoline-powered automobiles and the expansion of the U.S. federal highway system. Both events signaled a new era for the major oil companies: the more consumers drove their cars the more gasoline their cars consumed. This consumer need drove the development of the retail gasoline station; some historians place the first gasoline station in America in 1907 by Standard Oil of California (now Chevron) in Seattle, Washington.[1]

During the expansion years of the retail gasoline industry, notably, the 1940s and 1950s, the main points of brand differentiation between oil companies were the free services they offered their customers, including such items as trading stamps, maps, and windshield washing services. In more recent years, gasoline marketing and advertising has been primarily based on attracting customers with low prices, onsite car washing services, and the inclusion of food marts within the station. In marketing terms, this "lateral mapping" of adding several and distinct services into one location was an attempt by Big Oil's marketing executives to create a brand differentiator.

In an attempt to break out of the often vicious and unprofitable price wars that would oftentimes erupt from local retail gasoline competition, Mobil Oil launched several innovative marketing and advertising programs. The first of these involved the "advertorial" or advocacy editorial initiative that began in the early 1970s and ended following the Exxon acquisition of Mobil Oil. The second initiative was an in-depth market research project during the mid-1990s involving interviews with two thousand customers to gain new insights into the types of services gasoline customers wanted.

The advertorial initiative

On September 26, 1970, the *New York Times* newspaper launched its opinion-editorial (op-ed) page to guest columnists in its lower right quadrant as a place where organizations could pay for advertising. Mobil Oil took advantage of this unique opportunity and soon purchased advertising space in every Thursday edition, which it used for the next 30 years. Mobil sponsored its first op-ed page advertorial on October 19, 1970.

The purpose of the advertorial initiative was to inform numerous stakeholder groups about its operations and commitments to local communities, the energy industry, its contributions to the U.S. economy, and Mobil's position on a broad range of public policy issues. The company's advertorials (advertising-editorials) were designed to reach several highly influential stakeholder groups: the public at large, opinion leaders, and government decision makers.

In advertising jargon, advertorials are like all advertisements, paid messages in the media that are sponsored by a company or organization that seeks to tell its side of the story. Mobil's advertorial program was designed to give the company more public visibility and more importantly, to place the company at the center of national debates including energy, the environment, and international trade. The objectives of the initiative included swaying opinion leaders and the general public, playing a major role in the positive shift in attitude changes toward major oil companies, improving Mobil's image with the public, achieving the objective of building a reputation for Mobil as an outspoken, responsible company concerned about America's energy future and major social issues, and finally, having a favorable influence on investors. This last image building initiative was based on expanding the advertorial initiative during the mid- to late-1990s to include investor-focused advertorials in *Barron's* and the *Wall Street Journal*. These financially focused advertorials sought to make Mobil an investment of choice among other oil companies.

The strategy of the advertorial

In 1969, and before the energy shocks of the 1970s, Mobil was able to anticipate national and global energy shortages and rising oil and gasoline prices through several intelligence gathering operations: its internal news surveillance and information gathering and analysis processes which included political and economical intelligence compiled through its global network of subsidiaries. Mobil's leadership team quickly surmised that it and the other major oil companies would soon suffer negative publicity and media coverage. In an effort to tell its side of the story, Mobil used the op-ed page of the *New York Times* to begin a preemptive marketing and advertising campaign in order to defend itself against what it believed to be unfounded accusations and to voice its concerns over public policy positions while at the same time, advocate its own positions on public policy issues. Simply put, the Mobil initiative focused on two types of advertorials: image building and advocacy.

Mobil's CEO at the time, Rawleigh Warner, and Herbert Schmertz, then Mobil's vice president for public affairs, set three goals for the advertorial campaign:

- Build a reputation as an outspoken but responsible company concerned about America's economic and energy future
- Initiate and inform the public through a public debate on major issues affecting the company
- Broaden the spectrum of viewpoints available to Americans in high profile news media such as the *New York Times* that Mobil felt were antibusiness and specifically, antioil.

Advertorial content areas

In analyzing Mobil's advertorials over the 1985 to 2000 period, researchers Brown and Waltzer (2005) compartmentalized them into broad categories with a grand total of 819 advertorials. The top five leading categories of advertorials, numbers and percentages of the total, were as follows:

- International development and trade (85, 19.2%)
- Environment (84, 19.0%)
- Energy (71, 16%)
- Taxes (57, 12.9%)
- International affairs (22, 5.0%)

When viewed in hindsight, advertisements can often be viewed as a commentary on society and the world at large. For example, the Brown and Waltzer study looked at Mobil's advertorials from 1985 to 2000, which corresponds to the terms of U.S. presidents Reagan II (1985–1988), Bush I (1989–1992), Clinton I (1993–1996), and Clinton II (1997–2000). These presidential terms correspond to several tumultuous decades in world history including the ascendancy of Japan in world trade and economics, the fall of the Berlin Wall reflected in the "Wealth of Nations" advertorial series, future consequences of the 1990 Clean Air Act, and a multi-part advertorial series on global climate change. Today's

TABLE 12.1 *Marketing taxonomy at the gas pump: mobil's segmentation of consumers*

Road Warriors 16% of Consumers	True Blues 16% of Consumers	Generation F3 27% of Consumers	Homebodies 21% of Consumers	Price Driven 20% of Consumers
Generally higher-income middle-aged men who drive 25,000 to 50,000 miles a year, buy premium gas with a credit card, purchase sandwiches and drinks from the convenience store, and will sometimes wash their cars at the station carwash.	Usually men and women with moderate to high incomes, who are loyal to a brand and sometimes to a particular station, frequently buy premium gasoline and pay in cash.	Generation F3 (fuel, food, fast) motorists who are usually upward mobile men and women—half under 25 years of age—who are constantly on the go, drive a lot and snack heavily from the gasoline station's convenience store.	Usually housewives shuttle their children around during the day and use whatever gasoline station is based in town or along their route of travel.	Generally aren't loyal to either a brand or a particular station, and rarely buy the premium line of gasoline, frequently on tight budgets, efforts to woo them have been the basis of marketing strategies for years.

Source: Kevin Lane Keller. *"Building, Measuring, and Managing Brand Equity, 2nd edition."* p. 124.

oil companies can take a lesson from Mobil's initiative by adapting their marketing and advertising campaigns to current relevant themes.

From road warriors to homebodies

At the same time Mobil's advertorials were appearing in the *New York Times*, Mobil conducted in-depth market research surveys with a random sample of their customer base during the early- to mid-1990s. According to their research, only 20 percent of Mobil's customers bought gasoline based solely on price as opposed to brand loyalty of one particular brand. The market research surveys led the company to conclude that many consumers would forsake gasoline discounters in favor of a brand experience.

Specifically, Mobil's research revealed five primary purchasing groups labeled Road Warriors, True Blues, Generation F3 Drivers (for fuel, food, and fast), Homebodies, and Price Driven. Each of these groups exhibited different needs, personalities, and spending habits. The Price Driven group spent no more than US$700 annually, whereas the biggest spenders, the Road Warriors and True Blues, averaged at least US$1,200 a year. Mobil decided to target these big spenders, as well as Generation F3 Drivers, because Mobil felt many of them were destined to become Road Warriors.

Mobil selected Road Warriors as its target market and developed a brand positioning strategy it referred to as "Friendly Serve." Although retail gasoline prices rose by a few cents at Mobil stations, the company also committed new resources to improving all aspects of the gasoline-purchasing experience. Cleaner restrooms and better lighting alone yielded sales gains between 2 and 5 percent.[2] Additional station attendants were hired to run between the pump and snack bar to get Road Warriors in and out quickly—complete with their sandwich and beverage.[3] At the time the program began in 1995, station sales increased by another 15 to 20 percent. With a 10 percent share of the U.S. gasoline retail market during the mid-1990s, Mobil overtook Shell Oil in to become the largest gasoline refiner in the U.S.

The market research data Mobil compiled allowed them to make successful market introductions of several other retail products that further served to create a brand differentiator for the company. In 1996, Mobil introduced "Friendly Serve," a program that included elements of marketing campaigns from years past that included full service at no extra cost. Stations that decided to adopt the new level of service instructed its attendants to approach customers and offer to pump gas and wash windows free of charge. This program, designed to improve customer loyalty, enabled Mobil to increase prices. Today's gasoline retail environment and new competition from the likes of Wawa, Sheetz, and Costco, and including several grocery store chains, have placed even more pressure on the major oil companies to improve their consumer retail experience.

Mobil provided another convenience for consumers when it introduced the Mobil Speedpass technology in 1997. Speedpass picked up where credit card payments and pumps left off by enabling customers to pay electronically by waving their transponder, which drivers attached to their key chains, in front of the pump. Speedpass could also be used to pay for items from min-marts at Mobil stations.

Brand lesson no. 8

Mobil's advertorial program was innovative as it provided the company with a public forum designed to create its own unique brand. The company's advertorials were focused primarily on issues that would likely impact its operations and profitability, namely climate change, taxes, and other forms of government policy. But they were also informative as well. In one advertorial, Mobil described the rationale of taxes enacted by the U.S. federal government and the states on retail gasoline, a topic that many Americans previously knew little about. As oil companies move from print advertising to social media, the same lessons apply as they did earlier: focus on issues likely to impact operations and profitability; promote acts of community involvement; and inform and educate stakeholders.

Brand lesson no. 9

Mobil's market research surveys in the early- to mid-1990s served to demonstrate that traditional market segmentation still has a role to play in today's competitive retail gasoline market but little marketing innovation has been observed since then. Oil companies should take note of the advances made in consumer engagement from the likes of Pep Boys, an aftermarket retail chain of auto parts stores. Pep Boys provides wireless access and other non-traditional amenities so that consumers stay longer and spend more money. Wawa, Sheetz, and other retail outlets that provide a consumer experience rather than a simple pay-and-leave format are winning today's retail gasoline war. The gasoline station is the first line of engagement that consumers have with Big Oil. By becoming more consumer friendly, Big Oil can become more brand relevant.

Notes

1 http://en.wikipedia.org/wiki/Filling_station
2 O'Guinn, Allen, and Semenik. "Advertising and Integrated Brand Promotion, 6th edition." *South-western/Cengage Learning* (2012), p. 222.
3 Ibid.

13
The Brand Disconnect between BP and "Beyond Petroleum"

Abstract: *During the late 1990s and early 2000s, three of the world's energy giants, British Petroleum, Amoco, and ARCO, merged to become BP. Consolidating three organizations, their employees, cultures, and three brands was a major challenge. Getting the message out to the public that this was a new company with a new image was part of the marketing objective. Instead of focusing on selling retail products including gasoline, beer, soda, and snack foods, BP marketers created several new images of the company by launching "BP on the Street" and "beyond petroleum." These new images were directly opposed to BP's product line which was exploring and producing oil-related products.*

Robinson, Mark L. *Marketing Big Oil: Brand Lessons from the World's Largest Companies.* New York: Palgrave Macmillan, 2014. DOI: 10.1057/9781137388070.0020.

A Londoner named William Knox D'Arcy founded what eventually became British Petroleum in the late 1800s. D'Arcy made his fortune in Australia's gold rush and his plan was to become rich and live out a quiet retirement. But like most driven men of their day, being rich was simply not enough. The great industrialists and innovators of nineteenth century business, including John D. Rockefeller, Sr., J.P. Morgan, Andrew Carnegie, and Cornelius Vanderbilt craved greater success and, ultimately, more wealth. By the late 1880s, with increased oil discoveries and more derricks being constructed, oil was quickly becoming more important than coal, then the main source of energy of the day.

As recounted by Yergin (1991) to keep his fledgling company afloat, D'Arcy required more working capital, which was ultimately secured through the help of Burmah Oil, a British company whose founder was a Scottish millionaire who had made his wealth in Canadian railroads.[1] Burmah's investors, who were now in charge of drilling operations in Persia, now Iran, were successful in finding oil. This new discovery led to the formation of the Anglo-Persian Oil Company, the predecessor company to British Petroleum. Initially registered on April 14, 1909, as the Anglo-Persian Oil Company, Ltd., it was renamed the Anglo-Iranian Oil Company, Ltd., in 1935 and changed its name to the British Petroleum Company Limited in 1954.

At this point in its corporate existence, BP required additional financing for drilling. There were rumors that BP might be a takeover target for Shell, its main rival in the region. Following Winston Churchill's appointment as First Lord of the Admiralty, the civilian head of the Royal Navy in 1911, he recognized the strategic importance of the company remaining in British hands. Churchill's appointment to the position was pre-World War I, and many in the government were concerned about the looming threat of war with Germany. In a decision that would forever change the course of oil history, Churchill led the charge to convert the naval fleet from coal to oil. Based on its unique chemical properties, oil would allow the government to commission much faster ships giving it a competitive advantage during wartime. The changeover, however, from coal to oil had one major drawback: much of the British military's fuel was coming from foreign countries. In his remarks to the British Parliament, Churchill argued that to ensure success to his strategy, the government had to own or at least control the sources of oil that were to power its defenses. Rather than make purchases itself, the Admiralty paid £2 million for a controlling stake in Anglo-Persian. The deal required some unique changes in the corporate management structure, including that the company's directors be royal subjects. The government also appointed two directors with veto power.

While government funding allowed Anglo-Persian to keep operating, one of the impacts on the company was that it took on the organizational structure of a National Oil Company or NOC (many of today's largest non-publicly traded oil companies have retained the NOC-model of a government ownership structure including Saudi Aramco, Kuwait Petroleum, Petrobras, PetroEcuador, and Chinese National Offshore Oil Corporation). Historically, NOCs functioned as a government agency with all of its management and process inefficiencies (over the last several decades, however, NOCs have been improving both their management and technical capabilities to remain competitive in the oil industry).

From the early 1900s through the 1950s, Iran remained BP's primary source of oil. In one early example of the geopolitics involving oil, Iranians were becoming less complacent with control of their natural resource being given to a foreign company. With a new prime minister having been elected in 1951—Mohammad Mossadeq—BP's oil assets were nationalized (taken over and controlled by the government), the company was essentially kicked out of the country. The British government responded by implementing an embargo on Iranian oil and assembled a flotilla in the Persian Gulf. Historically, political influence over Iran had been exerted by the Russians and the British since the early 1900s, but now, British Intelligence, along with the U.S. Central Intelligence Agency (CIA), were concerned that the embargo would drive Mossadeq closer to the Soviet Union. In 1953, the CIA and the British Secret Intelligence Service organized a covert operation to overthrow Mossadeq as prime minister and restore the Shah as the ultimate ruler.

With the coup, BP was able to reclaim its Iranian assets and by 1954, the official name of the company became British Petroleum. Effective January 1, 1955, British Petroleum became a holding company. Beginning in 1977, the British government reduced its ownership of British Petroleum by selling shares to the public, and by the late 1980s the government turned over British Petroleum entirely to private ownership after selling its remaining shares. The name British Petroleum Company Plc was adopted in 1982. After merging with Amoco in 1998, the corporation took the name BP Amoco for a short time before assuming the name BP Plc in 2000.

BP on the street

Over a decade ago, three of the world's energy giants, British Petroleum, Amoco, and ARCO, merged to become BP. Consolidating three organizations, their employees, cultures, and three brands was a major challenge. Getting the message out to the public that this was a new company with a new image was part of the marketing objective.

As retail gasoline prices increased—at least up to the 2008–2009 financial crisis—the public image of oil companies in general had been poor. BP's marketers had a major challenge to overcome. The merger offered the perfect opportunity to create a fresh image in consumers' minds.

Instead of focusing on selling retail products including gasoline, beer, soda, and snack foods, BP marketers created a new image of the company by launching an ad campaign called "BP on the Street." "When you undertake an image campaign, it's critical that you know what you want to do," said Kathy Leech, BP's spokesperson at the time.[2] "We had two tasks. The first was informative. We had to let people know who BP was. The second was positioning. The goal was to lift BP from the negative aura that surrounds energy companies in the minds of the public. We positioned ourselves as a different kind of energy company."[3]

The tag line of the new campaign was catchy: "Beyond Petroleum." This conveyed the message that BP is more than a company that sells fossil fuels. As an energy company, BP wanted to convey the image to its target market that it was interested in doing more than fill consumers' gas tanks. BP wanted to provide answers to the tough questions concerning the energy industry. "It is the responsibility of an energy company to provide heat, light, and mobility to people. But you have to recognize that there are environmental costs," said Leech at the time.[4] "And you have a responsibility to mitigate those effects as much as possible."[5] At the time, BP was the first large energy company to acknowledge the existence of global warming and to take steps to reduce the impact of its operations on the environment. The firm invested in new sources of energy such as solar, wind, and hydrogen; research in climate change; and energy security—and these programs were featured in the initial advertising campaign, explaining to consumers why these activities are important to everyone.

One part of the challenge was to make people aware of the new BP brand as well as its name, while another part was to get them to relate to the message. So BP marketers created advertisements featuring real people voicing their concerns about energy issues. "The big issue in any kind of advertising is that people are cynical," admitted Kathy Leech at the time.[6] "They are especially cynical about oil companies. By using real people in the ads, and speaking in unscripted situations, we hoped to cut through some of that cynicism."[7]

The ads were initially launched in a few targeted major cities, including Chicago, New York, Washington, D.C., and London. The idea was to test some markets and observe how viewers responded to the message. Although consumers liked the underlying principle, the ads themselves were not popular. During the tests' first year, the feedback provided by consumers was that the ads were too negative. In general, consumers view oil industry marketing and advertising with a great deal of skepticism. So the BP marketing team set out to retune the ads. The new ads were provocative without being negative. The company sought a partnership with consumers. Rather than focusing on what BP had done previously and was currently doing, the new advertisements focused on what the company and consumers could achieve together. BP refined its ad campaign further, targeting an audience that it referred to as opinion leaders—those who vote, who follow decisions made by Congress, and who engage with their representatives via Internet blogging. Because advertisements can't reach everyone, BP targeted people who were more informed, to whom other people would go to for their information.

As the "BP on the Street" campaign moved from local advertising outlets to national and eventually international media, targeting its audience became even more important because it allowed BP to better monitor its advertising costs. The cost of advertising rises tremendously as media outlets expand. Despite this expansion, however, BP made local refinements wherever necessary. Leech noted that American consumers are receptive to British accents, but British consumers don't respond well to American speakers in commercials. German consumers don't care for "person on the street" ads, so BP created an "expert on the street."

Because of the high cost of an advertising campaign, and because conveying the right message is so crucial, BP marketers tracked the progress of "BP on the Street" carefully. "We tracked what was called a key learning summary at the end of each quarter," said Leech. "We made adjustments based on what we learned at the time. For example, in one year, we found that we presented too many messages. So we scaled back."[8]

Moving beyond BP

Moving beyond petroleum was essential for an energy firm like BP as it competes in the twenty-first century. As the firm was transformed to meet new challenges, it also changed the messages it transmitted to the public.

One marketing strategy that companies execute to distinguish themselves from their competitors is to create a brand differentiator(s). The differentiators that BP used beginning in 2000 were initially very successful. To boost its new brand commitment to the environment and renewable energy, BP invested heavily in both solar and wind power, leading it to alter its corporate identity, from British Petroleum to "beyond petroleum." The lower case letters were intended to signify a softer, more responsible corporate citizen that looked to a long-term future built on sustainable energy resources. Second, the company's logo changed to a yellow and green sunburst to reflect this new eco-friendly brand positioning. Both brand differentiators were brilliant and initially successful.

The merger wars and corporate name changes

In 1998, BP acquired Amoco Corporation of the United States and within the next 18 months had also acquired another U.S. rival, Atlantic Richfield Company (ARCO) as well as Britain's Burmah Castrol Plc. This spending spree reinvigorated a dormant industry and soon more M&A (mergers & acquisitions) transactions were being completed. Following years of rumors, Exxon and Mobil agreed to merge in late 1998 and by the end of 1999, the resulting merger led to the formation of the world's largest publicly traded oil company, Exxon Mobil. Just two years later, in 2001, Chevron and Texaco merged, becoming for a short time, Chevron-Texaco, and finally Chevron. The US$18 billion merger between Conoco and Phillips Petroleum formed ConocoPhillips.

By making these acquisitions, BP was literally transformed overnight from what was once a dormant national oil company into a Fortune Global 10 company. While making the company twice as big as it was just two years before, and allowing the company to better compete against the largest NOCs, the acquisitions had serious ramifications on company management. Government agencies are not known for their quick decision-making skills and it was the inheritance of the British government's ownership and management structure that played a role in many of BP's future catastrophes. Further, the ability to incorporate the personnel of Amoco, ARCO, and Burmah Castrol into BP did not allow post-merger integration projects to be fully incorporated, thereby placing the company at a competitive disadvantage in the marketplace.

The greening of BP

Following BP's acquisition of Amoco in 1998 for US$57 billion, CEO John Browne set out to reposition the company as something more relevant. The company strategized with corporate image consultants, Landor & Associates, and changed the company's name from British Petroleum to BP. At the time, BP's advertising agency, Ogilvy & Mather, began using the appropriate tagline "beyond petroleum," which the agency described as allowing BP "to reinvent itself as an energy company people can have faith in and inspire a campaign that gives voice to people's concerns, while providing evidence of BP's commitment, if not all the answers."[9]

BP's green rebranding efforts officially began with the announcement of its new "BP Helios" mark, named after the Greek sun god. The new logo did away with 70 years of corporate branding, replacing the BP shield, long associated in consumers' minds with the strength of British imperialism. The Helios mark cost US$7 million to develop and was forecast to cost the company another US$100 million a year to integrate into marketing, advertising, and company operations. At the logo's unveiling, the company's CEO, Lord Browne, directed attention to the company's recent purchase of the solar energy company Solarex, an acquisition that made BP the world's largest solar energy company. The unveiling of the Helios logo was a formalization of a rebranding strategy that had begun to emerge the year before with Lord Browne's announcement that 200 new BP sites around the world would be powered in part by solar energy, through solar panels placed on the roofs of gas pumps, and his commitment to reducing BP's own carbon dioxide emissions by 10 percent by the year 2010. From the outset, however, environmental groups heaped scorn on BP's green rebranding. Greenpeace gave the company its Greenhouse Greenwash Award, given to the largest "corporate climate culprit" on earth.

In contrast to the rest of the oil industry, and to further reinforce its brand, BP supported a number of environmental initiatives and organizations including the Kyoto Protocol and joined the Pew Trust's climate change program—the Business Environmental Leadership Council—that requires members to implement a voluntary carbon emissions reduction program. BP also joined several other highly influential groups including the International Climate Change Partnership, the World Resources Institute, the Energy and Biodiversity Initiative, the Papua Conservation Fund, and the Climate, Community, and Biodiversity Alliance. In addition, the BP Conservation Program provided conservation grants in 58 countries. BP also implemented an aggressive, voluntary CO_2 reduction program in cooperation with the environmental group Environmental Defense. To establish credibility with the public, Environmental Defense published audited inventories of the company's emissions. BP also used advanced product design to reduce pollution. In one example of its forward thinking, and five years before U.S. federal government regulations were to take effect, BP marketed a gasoline that contained 80 percent less sulfur than ordinary premium gasoline.

BP's new brand positioning strategy was initially successful but it ultimately failed due to three high profile events: the explosion at the Texas City, TX, refinery in 2005, the oil spill in the Gulf of Mexico in 2010, and the spill of oil in the Alaskan tundra.

Out with the new, in with the old

BP's financial commitment to renewable energy has waned significantly since 2011. The company's most recent annual report includes earnings solely in the upstream and downstream segments of its business. The *Financial Times* recently reported that BP had put its U.S. wind power unit up for sale.[10] If the sale goes forward, it would leave only BP's biofuels business— principally sugar cane ethanol in Brazil—and some research initiatives as remnants of its drive into alternative energy under Lord Browne, who had pledged to set the company on a course "beyond petroleum." BP had already abandoned other initiatives in renewable and low-carbon energy, withdrawing from the solar industry and dropping its investment in the development of carbon capture and storage technology.

Other examples of a change in direction include Bob Dudley's—the company's current CEO—comments to reporters in early 2013 that the company had "thrown in the towel on solar" after trying to make money at it for 35 years. One year before, the company cancelled plans to build a US$300 million advanced cellulosic ethanol plant in Florida, saying it could earn better financial returns elsewhere.

BP's dual corporate brands

When creating a new corporate brand, C-suite level executives should consider how the new image—called the Desired Identity—contrasts with the company's Actual Identity.[11] Following BP's merger with Amoco, it created a corporate entity that embraced both organizations' multiple identities. BP's new corporate brand positioning strategy emphasized its environmental activities and aspirations (Desired Identity). This strategy conflicted with the operational reality and the Actual Identity of the company.[12] BP was and is a company whose business segments are focused on oil exploration and production, oil refining, oil and gas distribution and storage, and gasoline/petrol retailing. As Greenpeace made known, it had reported that only 1 percent of BP's activities came from renewable resources. What is now clear is that BP did not change its core business to match the Desired Identity. Although the company had a stated corporate aim of being more environmentally friendly and invested small amounts in both wind and solar energy, the Desired Identity was simply what the company aspired to be, rather than what it actually was. Additionally, oil company executives should consider establishing corporate brand guidelines that are described as *Credible, Durable, Meaningful, Profitable,* and *Responsible.*[13] As such, a corporate brand positioning needs to be created with these guidelines in mind:

- *Credible:* reflects reality (grounded in the corporate identity—an entity's defining and differentiating characteristics)
- *Durable:* can be maintained over the long term
- *Meaningful:* valued by customers and stakeholders
- *Profitable:* of strategic value to the organization and—when shares are owned by investors—afford benefits to shareholders/stockholders
- *Responsible:* takes into account the company's external corporate responsibilities—societal, ethical, and environmental sustainability

Credible. BP's investments of not more than 1 percent of their total investments in renewable energy do not appear to establish the claim of a company whose long-term future lies in alternative forms of energy.

Durable. Was BP's new corporate brand positioning durable over the long term? In hindsight, the answer is "No." The new branding strategy was implemented in 2000 and within five years the explosion at the company's Texas City, Texas refinery, coupled with the announcements of several pipeline leaks in Alaska and accusations that BP traders attempted to manipulate the propane market in 2004, negated any positive impacts BP might have accrued. In fact, within the months following the refinery explosion, the media seized on reports from employees that the company had placed profits above human safety.

Meaningful. Stakeholders, including non-governmental environmental organizations such as Greenpeace, the Sierra Club, and the Environmental Defense Fund, were supportive of BP's branding initiative but there was a disconnect between stakeholders and the company's *shareholders*. In hindsight, the BP oil spill in the Gulf of Mexico indicated that both the broader stakeholder group and the company's shareholders should both be top-of-mind for company executives, and oftentimes the former will trump the latter. Moreover, in the aftermath of the Deepwater Horizon event, BP suspended dividend payments to shareholders.

Profitable. Most investors who invest in a company's stock are looking for two things: either long-term appreciable growth and/or dividends. Up until the Deepwater Horizon event, BP's shares on the New York Stock Exchange (NYSE) had steadily risen, as had its dividend yield. By comparison, renewable energy projects such as wind and solar are typically subsidized, meaning that governments provide some form of a production tax credit to encourage further investment in the industry.

Responsible. In 2006, BP revealed that it had uncovered a potentially dangerous amount of corrosion along a 16-mile feeder line into the Trans-Alaska pipeline in Prudhoe Bay. In response, BP said that it might have to shut down as much as 400,000 barrels of oil a day. Had the corrosion continued to be undetected, it could have resulted in a significant environmental disaster. Earlier in 2006, in another part of the same pipeline that BP was responsible for maintaining, a spill of 200,000 to 300,000 barrels of oil had been found, making it the largest oil spill ever on Alaska's North Slope. It was only when the U.S. government demanded that the company conduct a thorough inspection of the rest of the pipeline that the corrosion was discovered. Questions remain as to whether these actions are those of a responsible company.

Brand lesson no. 10

BP's "beyond petroleum" corporate brand positioning was a bold departure for an oil company and for a short time, a believable tag line. As early as May 1997, CEO Browne delivered a speech at Stanford University in which he said that global warming was a real problem and that BP and its fellow oil companies needed to both acknowledge that reality and begin dealing with it. He was the first oil company executive, and in hindsight, the only company executive, to take such a stand.

Commenting on his work in helping to create BP's advertising campaign, John Kenney, writing in an op-ed in the *New York Times*, said that, "in looking at it now, 'beyond petroleum' is just advertising. It's become mere marketing—perhaps it always was—instead of a genuine attempt to engage the public in the debate or a corporate rallying cry to change the paradigm."[14] And if there is one immutable law of marketing, it is that a company should practice internally what it is preaching externally.

Notes

1. Daniel Yergin. *The Prize: The Epic Quest for Oil, Money, and Power* (Free Press: New York 1991), p. 135.
2. David L. Kurtz. "Contemporary Marketing." (2009 edition) South-Western. Video Case 16.2 *BP: Beyond Petroleum*, pp. VC–19.
3. Gregory Solman. "BP: Coloring Public Opinion," *Adweek* (January 14, 2008).
4. David L. Kurtz. "Contemporary Marketing." (2009 edition) South-Western. Video Case 16.2 *BP: Beyond Petroleum*, pp. VC–19.
5. Ibid.
6. Ibid.
7. Ibid.
8. Ibid.
9. Joe Nocera. "Green Logo, But BP Is Old Oil," *New York Times* (August 12, 2006).
10. Ed Crooks. "BP Puts U.S. Wind Power Units Up for Sale," *Financial Times* (April 3, 2013).

11 John M.T. Balmer and Stephen A. Greyser. "Managing the Multiple Identities of the Corporation." *California Management Review* 44 (3) (Spring 2002), pp. 72–86.
12 Ibid.
13 John M.T. Balmer. "The BP Deepwater Horizon Debacle and Corporate Brand Exuberance." *Journal of Brand Management* 18 (2), pp. 97–104.
14 John Kenney. "Beyond Propaganda." *New York Times* (August 14, 2006).

14
Chevron and the Evolution of Human Energy

Abstract: *The 1970s was not a good decade for an oil company. The Arab Oil Embargo of 1973–1974 and the Iranian political turmoil in 1979 led to both shortages of gasoline and high oil and gasoline prices. Both events cast oil companies with a highly negative brand image in the eyes of consumers. In an effort to improve its image with customers, the Standard Oil Company of California (Standard/Chevron) undertook several rigorous corporate image advertising and research studies, beginning in the 1970s and today, culminating with "Human Energy" campaign and the adoption of the "We Agree" tagline.*

Robinson, Mark L. *Marketing Big Oil: Brand Lessons from the World's Largest Companies.* New York: Palgrave Macmillan, 2014. DOI: 10.1057/9781137388070.0021.

108 Marketing Big Oil

The 1970s was not a good decade to be an oil company. The Arab Oil Embargo of 1973–1974 and Iranian political turmoil in 1979 led to both shortages of gasoline and high prices. Both events cast oil companies in an unfavorable light in the eyes of consumers. In an effort to improve its image to its customers, the Standard Oil Company of California (Standard/Chevron) undertook several rigorous corporate image advertising and research studies, beginning in the 1970s which culminated in the "Human Energy" campaign and the adoption of the "We Agree" tagline.

Television research—the 1970s

As far back as the mid-1940s through the late 1950s, Standard/Chevron held favorable attitudes in the western states of the United States, their home markets.

TABLE 14.1 *Attitudes toward Standard Oil of California from 1944 to 1957*

Year	Like %	Neutral %	Dislike %
1944	49	31	20
1948	54	21	25
1951	51	38	11
1953	52	38	11
1954	59	34	7
1955	60	28	12
1957	44	43	13

Source: Lewis C. Winters. "Should You Advertise to Hostile Audiences?" *Journal of Advertising Research* 17(3), June 1977.

TABLE 14.2 *Attitudes toward Standard Oil of California before and during Arab Oil Embargo (Western states)*

Date	Favorable %	Neutral %	Unfavorable %	Index[b]
June 1973	26	58	16	-18
February 1974	41	11	48	-49

Notes: [a] Lewis C. Winters. "Should You Advertise to Hostile Audiences?" *Journal of Advertising Research* 17(3), June 1977. [b] Percent favorable minus percent unfavorable for standard compared to average of five other large corporations: GM, du Pont, GE, U.S. Steel, and Bank of America.
Source: Lewis C. Winters. "Should You Advertise to Hostile Audiences?" *Journal of Advertising Research* 17(3), June 1977.

DOI: 10.1057/9781137388070.0021

These favorability ratings remained unchanged for close to three decades until the Arab Oil Embargo. As the embargo gripped the United States and other Western economies, Standard/Chevron's favorability ratings began to decline. Four months after the Oil Embargo began, attitudes towards the company were polarized, that is, consumers either liked Standard/Chevron or they hated the company, when compared to these same five large companies. By February 1974, for the first time in more than 30 years, more consumers disliked the company than liked it.

What Standard/Chevron's survey results demonstrated is that when gasoline prices are low, favorable ratings remain high, and, by comparison, when gasoline prices are high, favorability is low. More to the point, if one oil company had a favorable/unfavorable rating, other companies in the industry will likely be rated the same.

In the face of this consumer hostility brought about by the Oil Embargo, Standard/Chevron set out to improve its image that culminated in its "Chevron at Work" advertising campaign. Chevron's executives hoped the campaign would help connect Chevron with consumers by illuminating what the company was doing to help solve the current (1973–1974) energy crisis.

In developing the new advertising campaign, the company asked consumers to participate in the McCollum-Spielman test. The test, developed by the market research and advertising firm McCollum-Spielman Worldwide, measures three outcomes of television commercials and advertisements:

- *Cut-through awareness:* viewers should be able to recall the advertisement 30 minutes after seeing it and correctly identify the product or company which sponsored the advertisement.
- *Main idea:* viewers should understand the intended message of the advertisement and be capable of describing the main idea.
- *Attitude change:* the advertisement should have a positive impact on the viewer's overall attitude toward the product or company.

Participants in the McCollum-Spielman test were taken to a central location and shown two pilot TV programs. Halfway into the first program, the program was paused and viewers were shown seven commercials one after the other. Some commercials lasted 30 seconds while others lasted 60 seconds. After seeing these commercials, viewers were asked to recall and correctly identify the commercials (brand managers refer to this phenomenon as "brand recall"). Of those seeing the 30-second

Chevron commercial, 47 percent correctly identified that fact. Of those viewers seeing the 60-second Chevron commercial, 54 percent correctly identified Chevron as the sponsor of the ad. These scores corresponded very closely to established McCollum-Spielman norms, 50 percent and 58 percent respectively.

Following this initial test, Chevron asked the viewers if the commercials had changed their feelings about the company. The net favorability rating—rating Chevron more favorably minus the percent who rated Chevron less favorably—was +17 percent for the 30-second ad and +15 percent for the 60-second commercial. These figures were well above the McCollum-Spielman norms of +7 percent and +12 percent respectively. The main conclusion drawn from this historical research was that even though consumers were at the time hostile towards the Standard/Chevron brand, television commercials were able to successfully persuade consumers to view the company in a favorable way.

Corporate image advertising and research—the 1980s

Following the end of the Arab Oil Embargo and the upheaval in Iran, favorable opinions of U.S. oil companies sank to 39 percent. This was the same rating reported in the 1980s following the political and economic upheavals in Iran. Clearly, each crisis had a major impact on the image of the oil industry; the all-time unfavorable rating was 34 percent and was reached during the Gulf War. However, Standard/Chevron's public opinion research showed that the company's overall image was consistently better than that of the industry. The Standard/Chevron image rose and then fell with the industry image. The company's research also revealed something quite surprising: the image of the worst oil company was more strongly correlated with the industry image than with the image of the best company. Based on the company's research, it appears that oil companies with a consumer perception of negative have a bigger impact on consumer's overall perceptions than good ones.

In early 1982, Chevron's public affairs department began a more systematic research exercise to understand the factors that shape the image of the oil industry and its companies. In the earlier research, the intent had been to see if consumers could recall and correctly identify the television commercials as being sponsored by Chevron. The new research took that intent one step further. Chevron now wanted to find out if there was a relationship between its product/retail marketing efforts and the company's overall image.

Before this research could begin, Chevron acquired Gulf Oil—one of the original Seven Sisters of the modern day oil industry—in 1984. At US$13.2 billion it was the largest merger and acquisition transaction in the U.S. corporate history. The merger presented Chevron with the unparalleled challenge of integrating the brand personalities of two giant oil companies. Chevron was widely known in the Southeast and Pacific regions of the United States while Gulf was the major brand in the Southwest region, especially in the state of Texas. Chevron's executive team decided that rather than market two brands of gasoline and other retail products, the company would market a single brand throughout the United States This decision had two major ramifications for the newly acquired assets from Gulf. First, Chevron would have to convert approximately 3000 Gulf stations to the Chevron name in Texas, Louisiana, Arkansas, and New Mexico, and second, to develop a new, all-encompassing corporate identity.

Chevron management viewed this challenge as an opportunity. Rather than simply change the identity symbols of the retail gasoline stations, here was a rare opportunity to introduce Chevron to its new customers and create a new personality for the brand. The company committed US$50 million to this demanding task.

Opinion research

Since the earlier research was conducted in the mid- to late-1970s and demonstrated high levels of success in improving its corporate image, Chevron public affairs executives set out to make their research more formal by establishing an annual consumer survey called the "Public Opinion Monitor."[1] This survey gathered data on Americans' overall attitudes towards Chevron, its competitors, and the entire oil industry. The survey asked respondents to rate the company on a number of corporate image attributes. As an example, the 1981–1982 survey asked people to rate Chevron on 16 attributes including contributing money toward the health, education, and social welfare needs of the community; showing concern for the public interest; paying its fair share of taxes; and making public statements that are truthful. Once the questions were finalized, the public affairs team agreed that, taken in total, the 16 attributes combined to form three broad corporate image categories useful for devising an advertising campaign:

1. Marketing and business conduct
2. Environmental responsibility
3. Corporate contributions

The corporate campaign

For the development of the advertising campaign, Chevron hired the J. Walter Thompson advertising agency to organize six independent creative teams to develop a range of advertising themes with an environmental focus. The creative teams brainstormed ideas that were eventually converted into six television commercial themes for testing:

- Go softly and gently
- Kids
- Water
- Friends
- Rabbits
- Color it

Just like the McCollum-Spielman tests of the 1970s, the new ads were also tested using the same methodology. While five of the six ads scored an average when compared to the McCollum-Spielman norms, one ad stood out: the Water ads had a significant better result than the others. This ad used an emotional soft tone and illustrated what Chevron was doing to protect the environment. This creative approach was adapted to develop a new corporate advertising campaign named *People Do*. Subsequent Chevron television ads ended using the voice-over: "Do people care about the environment? People Do."

The *People Do* advertising campaign was launched in California in 1985 with an annual budget of US$5 million. In order to determine the campaign's effectiveness, Chevron conducted what's known in the advertising industry as *media tracking research*. To accomplish this task, a sample of the target audience was surveyed before the campaign was screened on television to establish a benchmark and then tested again about a year later to gauge changes in the respondents' perception of Chevron, including their awareness of the brand, sponsor identification, favorable attitudes towards the brand, and likelihood of purchasing gasoline and other retail products from Chevron.

During the period from 1985 to 1986, Chevron's management teams' strategy focused on converting the newly acquired Gulf service stations to the Chevron brand. The executive team decided that the *People Do* ads could be used as a pre-conversion advertising campaign before the names on the Gulf gasoline stations were changed. One of Chevron's toughest tasks derived from research showing that Gulf customers had a different lifestyle profile—called psychographics—compared to the typical Chevron (California) customer, who valued security, conformity, tradition, and major established brands.

Research conducted during 1987 showed that some Gulf customers held strongly favorable images of the Gulf brand and that Gulf had become part of Texas history; this was ironic since the company had been headquartered in Pittsburgh for many years. In contrast to the Chevron customers who primarily valued security, conformity, tradition, and major established brands, Gulf customers were strongly motivated by loyalty to a perceived Texas institution. Thus, the marketing problems associated with converting the brand went beyond consumers' utilitarian needs of good service, convenient location, and low prices to include emotional ones like pride in Texas. One survey showed that 54 percent of Gulf customers would consider defecting if Gulf gasoline stations became Chevron stations (marketers refer to this as *brand switching*). Since the four states dominated by Gulf—Texas, Louisiana, Arkansas, and New Mexico—represented approximately 25 percent of Chevron's total U.S. sales volume, there could have been significant economic loss if vast numbers of Gulf customers deserted the re-branded stations.

With this knowledge, Chevron's main goal was to use corporate advertising to introduce the company to its newly acquired customers in order to retain them when the stations were rebranded. Because the psychographic profiles of the California and Texas consumer markets were different, the J. Walter Thompson advertising agency was asked to produce two new ads that might be suitable for the Texas market. McCollum-Spielman Worldwide benchmark tested these two ads against one of the *People Do* commercials that were running in California. The tests were undertaken in three U.S. cities: New Orleans, Dallas, and Houston.

The *People Do* ad tested was called *Eagle* as it showed a golden eagle preparing to land on a power line. The ad went on to say that had Chevron not installed a wooden platform on the power line, the eagle would have been killed. In terms of potential consumer awareness, *Eagle* scored slightly above average. The ad received above-average scores on descriptors such as "imaginative," "believable," and "interesting." More importantly, the commercial received very high persuasion scores, namely, a positive attitude shift of 34 percent. Further, three out of four pretest respondents said they would be more likely to buy Chevron gasoline after seeing the ad.

The *People Do* corporate advertising campaign began running in each conversion market approximately three months before any physical rebranding of the retail service stations took place. In California during the period 1985 to 1993, the *People Do* campaign registered a steady rise in brand awareness from 36 percent to 71 percent. The campaign was also effective in the Gulf region, improving consumer awareness of the Chevron-Gulf merger by 17 percent. The number of customers who said they would defect from the new Chevron stations fell from 54 percent to 4 percent by 1989. Two other indices of the campaign's effectiveness showed that by the early 1990s Chevron held nearly the top market share in the Houston market, and, perhaps more importantly, it was rated as the oil company that could be most trusted to handle an environmentally sensitive project in a responsible manner.[2]

Melding the Chevron and Texaco retail brands

Following the acquisition of Gulf and the resultant image rebranding process, Chevron and the other big oil companies entered a new wave of industry consolidation through numerous mergers & acquisitions (M&A) transactions. In the late 1990s, BP acquired Amoco, ARCO, and Burmah Castrol in short order, while Exxon acquired Mobil. Potential companies for acquisition were becoming few and far between. With the rise of the national oil companies or NOCs—those oil companies which are majority-owned by foreign governments owning huge oil reserves—it became evident that Big Oil's strategy would have to evolve to "acquire or fade away." Those companies that acquired survived; those that did not would likely not survive for long among the Big Oil ranks.

A new corporate oil giant was born when Chevron acquired Texaco on October 9, 2001, for US$38 billion, creating a new company worth US$117 billion. Once again, Chevron was faced with a brand dilemma:

the necessity of incorporating both brands into one new, coherent brand.

Starting in 2004, Chevron envisioned both brands' retail gasoline products as being premium, but serving two distinct demographic markets. In management's view, Chevron would focus on gasoline quality with a greater feminine bias towards safety and reliability, while the Texaco brand would focus on power and performance through its ties to NASCAR (National Association for Stock Car Auto Racing). Chevron's marketing strategy in North America became focused on back-office operations where everything behind the consumer would be exactly the same. All of the systems, all the processes, all of the support, the supply chain, all of the invoicing, everything the company supplied to their retailers would be exactly the same other than the consumer image, promotion, sponsorships, and advertising. This strategic move allowed the company to have limited incremental costs supporting both brands, while offering two premier brands to different demographics.

Although it is not uncommon for an oil company or gasoline retailer to carry multiple brands as part of its portfolio, it is unusual to have two premium offerings post-merger, with Exxon Mobil being the exception. Chevron products were sold on the value of the brand over price. This was based on data from the market research firm NPD Group that showed a level of brand loyalty for Chevron at 39.2 percent and Texaco at 31.9 percent, both of which were considerably higher than industry averages.

Chevron management then focused on changing the name of the company from ChevronTexaco to Chevron. The company's research highlighted three themes. It showed company management that the Chevron brand had the greatest resonance among non-governmental organizations (NGOs), the financial community, governments, and the people that give the company access to resources internationally—stakeholders that are of extreme importance in the upstream area of the business. The Texaco brand was extremely important and well known to retail and lubricants customers as a consumer-facing brand in almost 40 countries. The same applied to the Caltex brand. It became very clear that the company's identity was fragmented as ChevronTexaco. As ChevronTexaco, the name was too long and consumers shortened it to Chevron or Texaco or simply, "ChevTex." The company needed to create a unified image to its important stakeholder groups. According to its research, Chevron had the highest recognition and had a good corporate reputation which stood for a lot of things that were very important to those stakeholders.

The power of human energy

Following Chevron's renaming of ChevronTexaco to simply Chevron, the company looked for ways to break away from the oil company advertising clutter to position itself as the leading oil company. The independent advertising agency McGarryBowen developed a series of television commercials during the fall of 2007 called "Untapped Energy." These commercials were designed to highlight the debate over oil, energy, and the environment. The commercials showed images of Chevron's 58,000 "citizens"—its employees—who were named "the greatest sources of energy in the world." The commercials were then followed by an interactive game called "Energyville," designed to demonstrate the trade-offs between different energy sources for the fictional "Power City." Developed in partnership with the Economist Group, several hundred thousand people in 170 countries played the game.

The launch of the new "We Agree" advertising campaign in the fall of 2010 was an effort to repair Chevron's tarnished image in the wake of a couple of high profile events: the Gulf of Mexico spill which impacted all oil industry brands and the multi-billion dollar lawsuit that claimed the oil company is responsible for oil pollution in Ecuador. These new ads included slogans such as "Oil companies should put their profits to good use" and "It's time oil companies get behind renewable energy," followed by the phrase "We Agree" in bold red letters. To build credibility for each of these print advertisements, Chevron included the signature of one of its executives along with the signature of a third-party organization executive providing additional financial and moral support to the issue being advertised.

Brand lesson no. 11

During Chevron's acquisition of Gulf, the *People Do* campaign built a new corporate image and provided the support from which the merged company could launch a rebranding campaign. The *People Do* advertising campaign contributed immensely to the pre- and post-conversion success by familiarizing customers with the new brand and its commitment to caring for the environment. In fact, without the company's investment in image and advertising research, it is possible that the brand conversion program would not have been as successful. The research that was conducted—from pretesting to advertising tracking—supports this hypothesis. Similarly, when Chevron acquired Texaco, the company was again presented with an opportunity to create a new image. Building on the success of the *People Do* campaign, Chevron employed the "Human Energy" campaign and used the "We Agree" tagline to remind customers of the company's people-focused approach to business. The Chevron example demonstrates that brands with a long history—*historical brands*—can evolve, but to be successful, brand managers must present a consistent corporate profile over time.

Notes

1 Grahame Dowling. *"Creating Corporate Reputations: Identity, Image, and Performance."* (Oxford University Press: London, 2000), p. 153.
2 Lewis C. Winters. "Does It Pay to Advertise to Hostile Audiences Using Corporate Advertising?" *Journal of Advertising Research* 27 (3) (1988).

15
Brand Building at Shell Oil

Abstract: *Royal Dutch Shell was founded in 1907 following the merger between a Dutch oil company, the Royal Dutch Petroleum Company, and a UK company, Shell Transport and Trading Plc. Instead of forming a single merged entity, Royal Dutch owned a 60 percent ownership share in the merged company, while Shell owned a 40 percent share. In the decades prior to the name change, Shell's marketing and advertising activity was typical of the majority of big oil companies; all of them were known for their technical expertise in the areas of exploration and production (E&P) and research and development (R&D) but none was renowned as being innovative, especially in the area of traditional marketing, brand management, and communications.*

Robinson, Mark L. *Marketing Big Oil: Brand Lessons from the World's Largest Companies.* New York: Palgrave Macmillan, 2014. DOI: 10.1057/9781137388070.0022.

Shell was founded in 1907 following the merger between a Dutch oil company, the Royal Dutch Petroleum Company, and a UK company, Shell Transport and Trading Plc. Instead of forming a single merged entity as in most merger and acquisition transactions, Royal Dutch owned a 60 percent ownership share in the merged company, while Shell owned a 40 percent share. The resulting company, Royal Dutch/Shell Group had two boards of directors where one board is common. Under Dutch law, Royal Dutch had a two-tier structure with a supervisory board and a management board while Shell Transport operated under the British governance system with a unitary board. This structure remained in place for close to 100 years after which the company took on a new name following the previously discussed reserves scandal in January 2004—Royal Dutch Shell Plc.

In the decades prior to the name change, Shell's marketing and advertising activity was typical of the majority of big oil companies; all of them were known for their technical expertise in the areas of exploration, drilling, and production (ED&P) and research and development (R&D) but none was renowned as being innovative, especially in the area of traditional marketing, brand management, and communications. This view is supported by the fact that, beginning in 1996, few brand experts would have noticed an oil company among a list of great brands or companies revered for their marketing prowess. A 1996 survey of the worlds' most important company brands did in fact show Shell at number 64, BP at number 70, while Exxon failed to appear in the annual list. An internal survey conducted by Ad Age International for Shell a year earlier showed that Shell's advertising spend on a global basis was significantly less than other global non-oil company brands including McDonald's (15 percent) and Ford (10 percent). Shell's spending on integrated marketing communications was average for the oil industry.

Complicating Shell's marketing strategy was its two-tiered ownership structure. In a large, multinational organization such as Shell's, which includes several hundred-country subsidiaries, each country was allocated a budgeted amount of marketing and advertising funds to spend on local marketing initiatives. These initiatives focused on local country competition but often required separate advertising agencies using different brand messaging strategies.

The Shell "Answer Man"

Shell's "Answer Man" was one of the most recognizable and successful of all of the oil company advertising campaigns. Beginning in the mid-1960s and running through to the 1990s, the "answers" campaign dealt not with gasoline or Shell's other retail products as one might expect, but rather on common questions from motorists including vehicle maintenance, repair and safety issues, and ensuring proper tire inflation as a way to improve fuel economy. In one example from the campaign, a motorist's car runs off a washed-out road during a storm and ends up in a mud-choked river 20 feet below. The driver later revealed that he survived because he remembered a Shell Oil television ad. Over the duration of the campaign, Shell freely distributed close to 1 billion booklets from its gasoline stations and the mail. Shell deemed the campaign highly successful, saying it had received 5,000 unsolicited "thank-yous" via email, letters, and phone calls from the public. A follow-up campaign consisted of the Shell Answer Books that were inserted into popular magazines and a toll-free telephone number so that consumers could receive advice from a live attendant.

From the mid- to late 1990s, all oil companies including Shell were being impacted by environmental crises and corporate social environment issues. In Shell's case, the company was still being vilified for its poor decision to dump the Brent Spar oil storage platform in the North Sea and for its failure to defend the writer Ken Saro-Wiwa from execution by the Nigerian government for protesting over Shell's oil exploration in the country. Clearly, Shell needed an image refresh.

Not only had the "Shell Answer Man" run its course and was retired, it was also out-of-touch with the new realities of the 1990s including sustainable energy, human rights, ethics, and community issues. In its place came a new advertising campaign stressing Shell's ethical principles titled "Profits and Principles: Is There a Choice?"

Profits and principles

The new ad campaign took its theme from a report by the same name, which Shell published in 1998. Headlines in the report and in the ad campaign included "Exploit or Explore," "Protect Endangered Species. Or Become One," and "Commodity. Or Community?" As in many of its other advertising campaigns, Shell relied on the advice of the J. Walter Thompson (JWT) agency, but it went one step further. A public relations (PR) effort included stakeholder discussion forums that addressed these issues in a face-to-face format. Shell executives were quick to point out that this was not an advertising campaign in the traditional sense but was a means of inviting debate on the issues for which Shell had been criticized. Mark Wade, then a member of Shell's social accountability team, said that the company wanted to communicate the message that "while we haven't got all the answers and we are not squeaky clean, we are not evil monsters."[1]

Anti-Shell protesters remained skeptical of the fact that taking an environmental stance may not necessarily improve the brand in the eyes of consumers. People's trust in oil companies has remained low for many decades, they pointed out, and campaigns such as this were unlikely to change public opinion. Rather than spending large amounts of money on marketing, advertising, and public relations, many Shell haters believed the company should have invested in changing their operations.

A further attribute of the "Profits and Principles" campaign was a new focus on placing Shell employees at the center of its corporate marketing strategy. In conjunction with Shell, J. Walter Thompson created three 60-second television ads featuring the beliefs and attitudes of a Shell employee. The ads featured an individual's job and the contribution they make to the world on behalf of Shell. The initial three ads featured a geologist, a developer on a rural electrification project and a researcher working on a project to develop hydrogen as a source of fuel. In explaining its corporate marketing strategy revamp, Mark Moody-Stuart, then Shell's chairman of the committee of managing directors of Shell, said: "The television phase is the start of showing Shell people living our values. These are just three Shell people, but there are another 100,000 with stories to tell."[2] But in response, Body Shop founder Anita Roddick attacked Shell's ethical image campaign by remarking: "I trust J. Walter Thompson's new £15 million 'ethical' ad campaign will feature Shell staff in Nigeria. Then they can tell us why the Ogoni people still lack even basic social and environmental amenities when their land has produced millions of pounds in profits for Shell over the last 40 years."[3]

Count on Shell Oil

To raise its profile as a consumer-focused gasoline marketer in the United States, Shell Oil launched a US$50 million advertising campaign entitled "Count on Shell." The campaign was the most expensive marketing program in the company's history and is an example of public education advertising. "Count on Shell" attempted to position the company as customer friendly and knowledgeable about important issues facing motorists.

For this campaign, Shell developed print and television advertising that contained driving safety information reinforced by dramatic scenes and illustrations. In one example, a television ad demonstrated how a driver could avoid being trapped underwater in a car. The ad explained that if the driver could not get out through the window, he or she must wait until the car fills with water and the pressure on both sides of the door is equalized before opening the door and exiting the vehicle. Another ad instructed viewers to avoid braking while the voiceover intones, "In the blink of an eye, this man will have to resist a basic human impulse. Can you?"

The company supplemented its television ads by distributing 30 million free instructional guides called "Driving Dangers." One such pamphlet was entitled "Crash Course" and contained information on what motorists can do to help others in the event of a crash. Joe Kilgore, at the time, executive director of Ogilvy & Mather's Houston offices—Shell Oil's agency for the campaign—explained that the "Count on Shell" campaign reconnected with traditional Shell advertising: "People remember Shell as the kind of company that used to give them a lot of information. Shell had a foothold there, so it decided to take advantage of it."[4]

In the year following the "Count on Shell" launch, Shell received more than 5,000 emails, letters, and phone calls from consumers for whom the educational advertising worked. One woman detailed how Shell's advice helped her avoid a crash when her tire blew out on an icy Alaskan road.

Another important aspect of the "Count on Shell" campaign was a series of ads that touted Shell's technological advances in areas other than fuel. An ad featuring a professional snowboarder demonstrating tricks explained how Shell manufactures resins used in the making of a snowboard. Another ad showed a baby in diapers and conveyed the message that Shell polymers hold the diaper together. "This corporate strategy allows us to let more people know about the benefits Shell is providing," said Ogilvy & Mather's Kilgore.[5] "The company wants to put a lot of information out there so people won't see Shell as a one-dimensional company."[6]

The "Count on Shell" campaign had a positive effect on earnings from the sale of Shell Oil products in the United States, which rose to US$264 million in 2000 from US$98 million the previous year. Following the campaign's success in the United States, Shell exported it to 12 other countries.

A more recent campaign included the development of a series of television, print, online, and outdoor ads to tout the company's efforts to "unlock" cleaner sources of energy and let the world know that Shell is "ready to help tackle the challenge of the new energy future."

Focus on the fuel...

At the turn of the 21st century, Shell began to lose sight of its primary role as a provider of fuel. In the previous decade, the company spent its marketing and advertising dollars primarily on dealing with ethics, climate change, and human rights. Beginning in 2003, Shell, along with its primary ad agency J. Walter Thompson, began a return to fuels marketing. As reported by *Adweek* magazine, a new US$25 million advertising campaign made a value appeal with the claim that Shell's brand of gasoline offers better mileage than its competitors. This new ad campaign included television, print, radio, direct marketing, public relations, and point-of-purchase components.

In a television spot, three cars are identical in every attribute except in color. While driving, two of the three cars eventually run out of gas and pull off on the side of the road. The remaining car—with a subtle Shell logo on the side door—continues on to its destination. A voiceover says, "One of these cars is filled with new Shell gasoline. A gasoline specially formulated to give you better mileage. In fact, we have tests to prove it. So, if you took these three cars on an identical journey, the new Shell gasoline would help one of them go farther."

The television commercial was Shell's first fuel push in more than 10 years and was positioned as an engine-cleaning agent that also enhanced performance. The refocus on fuel was made because "consumers were getting more price-conscious, and they were looking for a good value," according to Karen Wildman, manager of national advertising and brands at Shell. "Getting consumers to focus on our quality fuel was one way we could compete."[7]

Oil companies spent a sizeable portion of their promotion and advertising funds during the 1990s on corporate citizenship initiatives. In doing so, "companies took notice that brand loyalty was beginning to diminish," according to Dan Gilligan, president of the Petroleum Marketers Association.

...and be passionate

Is it possible to be passionate about the brand of gasoline we put in our cars? Not very likely, but according to Shell, it is. Beginning on March 5, 2007, Shell began a hugely successful integrated marketing campaign that was designed to prove that all gasolines are not the same. The campaign entitled "Shell Passionate Experts" was created to demonstrate Shell's passion for what it does, its commitment to fuel quality testing, and its desire to prevent gunky engine buildup.

As part of the campaign, two new television spots were created by J. Walter Thompson to reinforce Shell's consumer products theme "Made to Move." In one of the spots, NASCAR driver Kevin Harvick and two actors playing Shell lab engineers encourage motorists to use the company's gasoline so as to prevent critical engine parts. The second television commercial features actual Shell employees and the passion they have in proving that all gasolines are the same. Both television commercials were seen on national network and cable channels such as ESPN, CNN, Discovery, and E! Print outlets including *Sports Illustrated*, *Golf Digest*, *Popular Mechanics*, *Motor Trend*, and *NASCAR Scene*.

Shell rehabilitates a tainted image

What differentiates the Shell brand when compared to its Big Oil brethren is its commitment to venturing out into the public arena to address its critics head on. "As we started considering the idea of a speaking tour, one of our first questions was, 'How will we be able to measure the effectiveness of face-to-face outreach,?'" explained Chris Bozman, deputy director of U.S. communications for Shell Oil Company.[8] "We really used research strategically to drive the campaign," Bozman says.[9] "We used it to understand audiences' perceptions about the energy industry, energy policies, and Shell Oil Company and to measure the effectiveness of the communications on this issue. Then we used the data we gathered to shape ongoing messages about the need for public policy changes."[10] The result of Shell's strategizing was the development of a "National

Dialogue on Energy Security" led by the company's then U.S. head, John Hofmeister.

In the beginning of 2006, it was clear that Shell Oil executives were facing a reputational crisis in its major market, the United States. What led to the crisis were a series of high-profile events. First, economic growth in emerging markets led to increased levels of oil demand. Since oil is the primary feedstock in refining oil into gasoline, this in turn led to increases in retail products such as gasoline. Second, several hurricanes in late 2005 battered the U.S. Gulf coast—a major U.S. oil refining region—causing supply disruptions leading to additional increases in oil and retail gasoline prices. Shell's and the oil industry's image took a further beating when it was reported that some rogue station owners had increased their prices exponentially to take advantage of the crises. The result was (1) negative media coverage, (2) complaint letters from motorists, and (3) summons for Mr. Hofmeister to testify along with other oil company executives before Congressional leaders.

Congressional leaders are known for their knee-jerk reactions to situations which are likely to infuriate their constituents. In this case, Shell needed to promote its message to two of its most important target audiences: policy makers, both state and federal lawmakers, as well as federal energy regulators. These audiences, could at any moment, yield to public pressure and create and pass legislation that could make a challenging operating environment for oil companies.

Hofmeister was the guiding force behind the creation, development, and execution of the Dialogue, and personally committed to taking Shell's message to 50 cities across the U.S. The groundbreaking event began as simple speaking engagements but was expanded by adding a town hall event in each market, being led by Hofmeister, and including several other high-ranking Shell executives. Media interviews were also arranged with small groups of the local population, educators, customers, Shell employees, and other key stakeholders.

Following each of these town hall meetings, participants were surveyed to give their input on the three questions Shell was discussing:

▸ What should the United States be doing to increase domestic oil supply?
▸ What should the United States be doing as a nation to manage energy demand/consumption?
▸ What is your vision of the United States energy portfolio in the coming decade and beyond?

Responses were gathered, summarized, and presented to the audience at the end of each event. The responses were also transcribed and posted to Shell's U.S. website. When all of the town hall responses were tallied and analyzed, 62 percent of those who had attended the town hall events were favorable toward Shell, compared to 33 percent of the general public. Not surprisingly, other oil companies also received more favorable ratings from attendees, but Shell was rated the highest.

In addition to the town hall surveys, three tracking studies were conducted in conjunction with Burson-Marsteller, Shell's agency of record, and its affiliate Penn, Schoen & Berland (PS&B).

The first tracking study was conducted in October 2006, after only a few presentations had been given. Not surprisingly, awareness of the events was low. This low rating served as the baseline upon which future studies would be benchmarked.

A second tracking study was conducted in February 2007 following a four-month touring period: October, November 2006 and January, February 2007. This study showed significant improvement in image perceptions of Shell among those aware of the tour. It also provided Shell with additional feedback that the company used to adjust its messaging (for example, spending less time describing the issues and more time talking about what Shell was going to address them).

The third survey was conducted during November 2007. The research showed that among those of the tour, Shell had a favorability rating of 61 percent at the end of the tour, compared to a 48 percent rating at the beginning of the tour. Upon further analysis, Shell received the highest rankings of all major energy companies among business leaders as an "energy leader" (25% in November versus 16% in October 2006). First-place rankings on this attribute also rose among other audiences: from 15 percent to 20 percent among community leaders and from 5 percent to 25 percent among media.

The "Let's Go" campaign

In an era of social media marketing, oil companies have a unique opportunity to engage consumers in ways no one thought possible just 10 years ago. However, in doing so, oil companies not having planned this new consumer engagement process thoroughly may face criticism over their highly negative legacy. A case in point is Shell's most recent campaign titled "Let's Go," which was launched in June 2010 with the tagline "Let's Broaden the World's Energy Mix." The initial campaign was launched with a slickly produced television spot depicting scenes of efficient transportation methods designed to promote the ways in which Shell's natural gas provides climate-friendly utility around the world. Within days of its release, a video supposedly showing the launch of the company's new Arctic exploration program appeared on YouTube. The video, which appeared to be taken from a smartphone, showed a corporate boardroom party taking place. In one scene, the formally dressed crowd was shown celebrating with drinks poured from a fountain shaped like an oil-drilling rig. As the first guest approached the fountain, it malfunctioned and spewed a dark oil-type liquid all over the guests and several hapless Shell executives standing nearby. Not surprisingly, the video was a viral hit. It was later revealed that Greenpeace and the legendary corporate activists, the Yes Men, had made it as a spoof. By July, while roughly 3,000 people had watched Shell's original, more than half a million had watched the spoof.

Brand lesson no. 12

Shell's marketing and advertising journey over the last 50 years has demonstrated that a company in a highly negative industry can improve its image. The company's advertising campaigns, a good portion of which featured educating consumers on a variety of issues including vehicle safety and emergency situations, connected with consumers. When oil companies focus their efforts on providing real benefits to consumers, the brand value is increased. Oil companies need to not only say what they're doing to improving consumers' lives, but also prove by doing it.

Notes

1 Steve Hilton. "Squeaky Clean? Who, Us? Shell Ads Play Truth Game," *theguardian.com* (March 27, 1999).
2 Harriet Marsh. "Shell's New Ethical Ads to Focus on Employees," *Marketing Magazine* (September 23, 1999).
3 Conor Dignam. "Roddick Attacks Shell's Ethical Ads," *Marketing* (September 30, 1999), p. 1.

4 Steve Krajewski. "Olgivy & Mather Create More 'Answers' in $50 million Image Campaign for Shell Oil," *Adweek* (February 9, 1998), p. 5.
5 Ibid.
6 Ibid.
7 Mindy Charski. "New Shell Campaign Focuses on the Fuel," *Adweek* (June 30, 2003), p. 15.
8 "Fueled by Research and Measurement, Shell Oil Rehabilitates a Tainted Image," *PR News*. (October 20, 2008).
9 Ibid.
10 Ibid.

Part IV
Big Oil and the Era of Consumer Engagement

16
Building Brand Loyalty: Improving the Retail Fueling Experience

> **Abstract:** *For today's consumer, a stop at the local gasoline or petrol station can feel like paying a visit to the dentist: you know you have to go, but you'd rather avoid it. While gasoline prices vary by location, many retail stations remain filthy and unkempt, while their outward appearance is poorly maintained. In theory, improving the consumer experience should lead to higher revenues and improved brand loyalty. Gasoline retailers can create a more positive image to their customers by making small but effective changes to the station appearance and improving station attendants' customer service training.*
>
> Robinson, L. Mark. *Marketing Big Oil: Brand Lessons from the World's Largest Companies.* New York: Palgrave Macmillan, 2014. DOI: 10.1057/9781137388070.0024.

Purchasing gasoline or petrol from the local station is one of the most unpleasant purchasing experiences consumers have to make. Many stations are poorly lit, are often in disrepair, and customer service is nonexistent. These failures in customer service need to be corrected as the local fueling station provides oil companies with their only opportunity to engage with the consumer. By making improvements in their retail infrastructure and operations, oil companies should experience higher revenues and increases in brand loyalty.

Why is building brand loyalty so important to Big Oil? One reason is that the gasoline station is the first, and often the only direct contact these companies have with their consumer base. Second, as competition has heightened over the years, Big Oil now finds itself competing against not only other Big Oil brands and specialty retailers, but also low-cost companies such as Sheetz, Wawa, Costco, and a myriad of convenience and grocery stores.

In the early days of gasoline retailing, station owners would give away free items including maps, stamps, and other novelty items. This was an attempt to build brand loyalty between the station and the consumer. As brand loyalty increased, the hope was that consumers would spend more of their household budget at one station. As time went on, Big Oil spent millions in marketing and advertising, hoping they could convince consumers that their gasoline was better than that of their competitors. Unfortunately, gasoline is a commodity and oil companies must now create an engaging customer experience if they want to remain relevant.

In the 1980s, retail station owners began to offer self-serve or pay-at-the-pump options. While this decision was a way to bring down costs at the local station, it did little to create a consumer experience. Gone were the days of a friendly station attendant offering a free windshield wash or a free check of your engine's oil level. In its place was one attendant in a glass-protected booth that would sell you a pack of cigarettes, gum, or candy. During the mid-1980s and into the 1990s, oil companies began experimenting with pay-at-the-pump credit card readers in an effort to make gasoline purchasing more convenient for consumers. This innovation did make the payment process easier but in the long run, it was not a better consumer experience. Mobil's previously discussed disruptive innovation, the Speedpass product, was another innovation in payment convenience, but the company's stations remained as poorly maintained structures. The Friendly Serve program that Mobil developed in conjunction with the Speedpass was the one of the few efforts oil companies have made in recent years in an effort to improve the customer experience and build brand loyalty. So how can oil companies and gasoline retailers improve the consumer experience? Retailers such as Starbucks offer some constructive ideas.

The Starbucks model

When Starbucks began to offer its premiere products back in 1971, coffee consumption in the United States was on a decline. To offset this decline, most coffee retailers of the day started using cheaper coffee beans to compete on price, hoping increased sales volumes would increase revenues. The company's founders decided to experiment with imported coffee beans, which they could sell at a premium price.

Following 10 years of growth, Herbert Schultz was brought in to run the company by its founders. During a trip to Italy, Schultz discovered the culture of drinking coffee as a consumer experience. In his words, "it was an extension of people's front porch. It was an emotional experience."[1] What led to Starbucks financial success was not just the addition of free Wi-Fi, but the amenities that customers wanted most: a unique customer experience, driven by quality products, clean stores, and friendly service. These are the basic ingredients necessary to create a strong brand. As Starbucks implemented its branding strategy, it found that as consumers stayed longer within its stores, they spent more money.

When the Starbucks model is applied to the retail gasoline station, several themes become apparent. First, where Starbucks encourages consumers to stay, gasoline station owners prefer consumers to stay as short a time as possible. When customers enter the food mart—that portion of the store set aside for food products, soft drinks, and other items—there is rarely a friendly greeting or other acknowledgment that the customer exists. Fuel station owners need to create a reason for customers to stay longer and spend more.

Creating the customer-focused station

Visiting the local gasoline/petrol station need not be the dull experience it is today. There are a multitude of enhancements station owners can make to create a truly unique customer experience.

Offer a drive-through window for food and other items. Gasoline retailers should take a page from the playbook of fast food retailers by offering drive through service. This would involve either adding a covered enclosure to one side of the station or creating more internal space for a drive-through window so that consumers can purchase the stations' products without having to venture inside to the food mart.

Improve the freshness of the food selection. Customers want quality food choices they can purchase quickly. Improving the freshness of food items at a given station will require in-depth market research to determine the appropriate product selection and pricing strategy.

Customer service training for station personnel. Walmart is known for its official store "greeters," while Nordstrom is renowned for its attention to the customer. Cashiers and other gasoline station personnel should be required to dress appropriately and be educated on how to acknowledge customers as they enter and exit the food mart.

Offer a loyalty program. Many retailers offer some form of a loyalty program designed to build and retain their customer base. For example, for every nine fuel fill-ups, the tenth is free. This is similar to the programs offered by many food retailers or coffee houses.

Clean restrooms. This is a must, especially if the station wants to attract female consumers. Many customer-focused retailers clean their restroom facilities every hour.

Redecorate/repaint the station, inside and out. A little redecorating and repainting can go a long way to creating an inviting station.

Keep hand sanitizer containers full. This is another must activity, especially as fuel pumps, hoses, nozzles and nozzle handles are located outside the station. Ideally, the equipment should be cleaned several times a day using a strong disinfectant.

Install or make better use of flat screen monitors on top of pumps. Many gasoline stations have installed flat screen monitors on the top of their gasoline pumps. The monitors make sense in that while consumers are spending the three to five minutes pumping their gasoline, they can view sponsored advertisements. Companies should revisit the use of the screens to see if station store sales increase or if they are meeting other marketing metrics.

Improve the exterior lighting. External lighting is important in terms of customer safety.

Re-evaluate placing product inserts in credit card statements. It's rare when customers purchase the items advertised in the inserts within the customer's credit card statement. Instead, a personalized "thank you" insert would prove to be more useful in demonstrating the company's appreciation for customer's patronage.

Mine the data from customer purchases. Big Oil companies possess a wealth of valuable market research data from their back-office databases and social media that can reveal intricate patterns of consumer purchase behavior. Today's consumer focused businesses are already utilizing data analytics and other forms of Big Data to crunch the numbers in an effort to improve their marketing and advertising.

Fueling the future

During the late 1990s, a robotic gas-station pump had been rumored to be in development from several of the major oil companies including Shell, Mobil, and BP. Several prototypes were demonstrated around the world but the technology had not been perfected. Recently, Fuelmatics of Sweden and Husky, based in Pacific, Missouri, introduced the next iteration of this product innovation.

In the current version, the two companies developed an automated pump that uses visual sensors to locate a car's fuel door that can add gasoline without the driver having to leave their car. Consumers drive up to a gas station pump, select the fuel they want, and pay at their driver's side window with their credit card. The pump structure slides along a track to the proper position. A robotic arm opens the car's fueling door using a suction-cup attachment and then inserts the fuel nozzle. According to the companies, consumers need to replace their car's fuel cap with a US$9.00 fuel filler. Will consumers gravitate to a robotic gasoline pump if their local service station offered it or is it just another marketing gimmick to lure customers into the station? It is hard to say, although with every product innovation, there are some early adopters who might use the new pump. Having a robot pump your gasoline will initially be viewed by consumers as a novelty item but won't likely do much to improve the overall experience.

Brand lesson no. 13

Improving the Big Oil brand at the retail level will require some innovation in the areas of station redesign, product offerings, and consumer loyalty initiatives. Much of this can be accomplished by changes to the inner and outer station façade. The real challenge will involve creating a unique and exceptional consumer experience that can't be matched by Big Oil's competitors.

Note

1. Phillip Kotler and Kevin Lane Keller. *Marketing Management*, 14th edition (Pearson Education: New Jersey, 2009), p. 648.

17
Communicating with the Masses: Big Oil and Social Media

Abstract: *There is no doubt that social media is changing the way people communicate and connect with each other. Not only do social media sites like Facebook, Twitter, and YouTube facilitate people-to-people communication but they also play a role in how consumers communicate to, and about, a company's brand. Big Oil brands can improve their communication with the masses by implementing many of the recommendations from this chapter.*

Robinson, Mark L. *Marketing Big Oil: Brand Lessons from the World's Largest Companies.* New York: Palgrave Macmillan, 2014. DOI: 10.1057/9781137388070.0025.

There is no doubt that social media is changing the way people communicate and connect with each other. Not only do social media sites like Facebook, Twitter, and YouTube facilitate people-to-people communication but they also play a role in how consumers communicate to, and about, a company's brand. But there remain several intriguing questions concerning how Big Oil brands are using social media. First, in what ways are Big Oil brands using social media and second, how successfully are they in using it?

Social media has already become one of the major communications tools in the corporate playbook, especially in the area of crisis communications, although it should also be used during the course of normal business operations. The main purpose in using integrated marketing communications, including social media, during normal business operations is to make the brand stronger so that when a crisis does strike, the company has accumulated enough brand equity to help deflect some, if not most, of the negative media scrutiny.

During previous incidents like the Exxon *Valdez* and Shell's *Brent Spar*, there was no social media. The communication channels available to those companies were traditional media including television, newspapers, magazines, and radio. During April 2010's Gulf of Mexico saga, BP was the first oil company to venture into the unknown realm of using social media to combat a major crisis. In his evaluation of BP's effort, Richard Kerley, Senior Digital Strategist at Washington DC-based Levick Strategic Communications said, "They're playing by old rules. Dealing with a crisis has totally changed because of social media."[1] Kerley went on to comment further that companies need to act proactively when operations are normal rather than react when a crisis strikes. When planned and implemented flawlessly, a social media campaign can help get important messages out to stakeholders. In the area of crisis communication, it is becoming more apparent that companies will have to update their communications plan to include best practices in social media. If executed poorly, social media can inflict further damage to the company's reputation and brand.

Communicating the Gulf of Mexico story

In confronting the Gulf Oil spill, BP successfully avoided many of the communications failures that impacted Exxon's ability to cope with the

Valdez incident. BP immediately dispatched its senior executives to the scene and provided regular updates as events unfolded. By contrast, Exxon's first senior executive didn't arrive on the scene until almost a week later. Moreover, BP took full responsibility for the incident almost immediately through press conferences and media outlets. The company also brought in employees from around its international offices with particular skillsets to support the media response effort.

Conflicted and separate communications

Whenever there is an incident of BP's severity and magnitude, several groups of stakeholders are immediately impacted. These include the company and its employees, the public-at-large, local, state, and federal government agencies, and environmental and other non-governmental organizations. Initially, BP handled press relations and media statements on its own, but as the situation progressed, the company was forced to include the U.S. Coast Guard in its daily spill evaluation and media responses. At the outset of the crisis, the U.S. Coast Guard's Rear Admiral and Deputy On-Scene Commander Mary Landry stated that there was little to no oil leaking from the Macondo well but days later, she acknowledged there was oil in the water which appeared to be seeping. Flow estimates then increased from an initial rate of 1,000 barrels per day (b/d), then 5,000 and then 20,000 by May. The rate was then revised upwards to 30,000 and then finally, to between 40,000 and 60,000 b/d.[2] Continuous reported revisions of spilled oil made it appear that no one, neither the company nor the government, knew the real facts. The situation can be made exponentially worse when using social media in a reactive rather than a proactive mode.

Weeks into the incident, media reports began to surface questioning who was in charge of briefing the public on the progress of the spill and the cleanup. On June 1, the 41st day after the accident, it was announced that Thad Allen, the national incident commander, would begin daily press briefings without BP. This action was seen by many as a way to ensure not only timely and more consistent delivery of information, but more importantly, to show that the U.S. government was actively involved in managing the response. As the spokesperson Thad Allen exuded the presence of a take-charge individual who possessed the skills necessary to connect with the public. He was confident, articulate, and had a solid grasp of the facts. He exhibited the image that Tony Hayward couldn't.

Can we advertise our way out of this?

During the event, BP undertook an integrated marketing communications campaign encompassing television, print advertising, digital/online (web), and social media. Social media provided BP with direct and immediate contact to a large audience that was previously unreachable. The television and print advertising campaign was well executed and the copy was focused squarely on what the company was doing to contain the spill. By mid-May, BP's advertising began to focus on employees who were engaged in the cleanup effort, pay claims, and save wildlife. The spokespeople were employees who lived and worked in Louisiana and other Gulf Coast states. In hindsight, this proved to be an effective strategy. By using local Gulf employees as the main spokespersons, BP provided a more human element to the media relations campaign that previously had been lacking. The public could easily identify with, and show empathy towards, citizens of the Gulf coast rather than an arrogant and aloof CEO wearing expensive business suits.

BP also provided extensive information updates on its web site devoted to tracking the progress of the response. The site was easy to navigate and informative, including updates on claims information, pictures, videos, and BP contacts. Visitors to the site had an option to receive email updates and view other content via Facebook, Twitter, Flickr, and YouTube. The entire media campaign effectively positioned BP as being a transparent company, dedicated to sharing information with the public.

As attempts were being made to plug the leaking well, BP initiated its marketing and advertising program, which was designed to counteract the negative media scrutiny it was receiving and to tell its side of the story. The company ran advertisements in high-profile media including the *Wall Street Journal* with initial advertisements describing the company's progress in plugging the well. Following the completion of this task, the company was forced to vacate this initial strategy in favor of attacking several other issues that had begun to arise: fraudulent insurance claims from local businesses, court battles over the executor of the compensation fund, and finally, the law firms who were also being overly compensated for their work on the insurance claims.

BP enters the Twitter sphere

Prior to the incident, BP had an active Twitter account but was not making regular communications to stakeholders. The official @BP_America Twitter account was launched on January 14, 2010, and only 12 tweets were sent prior to the spill.[3] Before the April 2010 incident, BP used Twitter sparingly, mostly to disseminate corporate information, but within weeks following the explosion, Twitter's use became a major strategic communications tool. BP launched a secondary Twitter account on April 29, 2010, specifically dedicated to its crisis response in the Gulf (@Oil_Spill_2010).[4]

BP transmitted a total of 1,142 tweets via its @Oil_Spill_2010 account (later changed to @Restore_TheGulf) between April 29, 2010 and September 19, 2010 which was the official date of the plugging of the well. Tweets were initially sent from BP's Twitter account, but eventually a "Unified Command" emerged which combined BP's effort with those of various government agencies such as the Environmental Protection Agency, U.S. Coast Guard, and a host of local and state agencies along the Gulf Coast. The total number of tweets sent out during the event from Day 1 through Day 142 totaled 1,142 and are detailed in Table 17.1.

TABLE 17.1 *BP's tweets during the Gulf of Mexico spill*

Days	Number of Tweets
1–15	180
16–30	464
31–45	125
46–60	59
61–75	76
76–90	95
91–105	43
106–120	45
121–135	45
136–142	10
Total Tweets	1,142

Source: Walton, Cooley, Nicholson. "A Great Day for Oiled Pelicans," BP, Twitter, and the Deep Water Horizon Crisis Response, (2012). *International Public Relations Research Conference*. Accessed (January 29, 2014), http://www.instituteforpr.org/downloads/690

BP's Twitter campaign

BP's initial tweets during the Gulf disaster indicate that its Twitter strategy was largely unplanned and reactive, as the company experimented with various communications channels to distribute information to its stakeholder groups. Based on the number of tweets in the first few weeks after the explosion, the company's strategy appeared to be focused on tweeting informational updates related to the spill. Three weeks after the incident, the company severed the link between its Facebook and Twitter accounts marking a significant shift in both message strategy and its use of available communication channels.

One month after the event, the most significant adjustment in BP's social media message strategy with Twitter accompanied the launch of its new blog. The company encouraged stakeholders to use the new blog as the platform for lodging complaints about the incident, rather than having its Twitter followers tweeting directly to BP's Twitter handle. The loss of this two-way form of communications was another major shift in message strategy. By July 2010, the U.S. government took control of the Twitter account and the name was changed to @Restore_TheGulf. From this point on there was no further two-directional communication recorded in tweets; no explanation was given by either the company or the government. Due to the shift in control of the account, the study conducted by Walton, Cooley, and Nicholson (2012) identified the U.S. federal government as the primary spokesperson for 50.44 percent of the tweets and BP or its representatives for 46.41 percent.[5]

In hindsight, much of BP's use of Twitter can be described as reactive and chaotic. The seriousness of the incident, combined with using a new media channel, was likely to be responsible for much of the chaotic atmosphere surrounding the event. BP also faced comparisons to other high-profile events such as the Exxon *Valdez* and the failure of the U.S. government in its disaster response to Hurricane Katrina. Early in its brand crisis, BP projected that it could be at least six months before the leaking well could be fully contained and plugged. With this viewpoint in mind, BP's management strategy can be viewed as simply weathering the initial brand attack. The shifting of BP's various social media strategies could be interpreted as a way to ensure the public was aware of the company's efforts to plug the well. Or was the company simply using social media to show that it was actively involved in the incident until such time as the company could figure out exactly what its messaging strategy was?

Creating brand engagement on Facebook and Twitter

Over the last five years, brands have actively embraced Facebook and Twitter as a way to create brand engagement with consumers as part of normal business operations. By using this new marketing channel, brands have been shown to drive engagement and to further increase brand awareness. In one recent study, several academics and a chief marketing officer from a consulting firm collaborated to code more than 1,000 wall posts from 98 global brands in an effort to understand how different wall-post attributes impact the number of Facebook "likes," "comments" and "shares" a post receives.[6] The findings of their research have been paraphrased into eight points that are worthy of consideration for Big Oil brands.

Photos matter. As the saying goes, "A picture is worth a thousand words," and so it is with sharing photos on Facebook. Placing photos from a company's product line on Facebook will likely elicit a number of "likes."

Stay topical. Although it seems obvious, keeping up with the times is important. Messages considered to be topical include anniversaries and important firsts in a company's history.

Continue to promote the company brand and its products. It is to be expected that when consumers visit a company's wall, product promotions are front and center. Consumers visit the walls of brands they identify with and want to engage with. When consumers actively engage in helping to promote a company's branded products, marketers call them "brand ambassadors," a form of consumer-generated marketing.

Share success stories. Consumers want to align themselves with brands they and their network identify with. Brands should make consumers aware of their awards and achievements from third-party evaluators.

Educate consumers. Brands should create informational value to consumers. The educational effort should include informing their followers on the history of the brand(s) and the value they bring to society.

Make the brand more human. Brands need to inject human emotion to position themselves as living objects. Companies that share posts containing emotion help fans identify more closely with the companies' products and services.

Humor can act as the best social medicine. People like to laugh and brands that inject humor into their wall posts receive a significant higher number of "likes" than brands that don't.

Ask to be "liked." Brands that asked to be "liked" on Facebook tend to see their "likes" increase. Granted, this should be done in a non-sales manner.

Getting messages retweeted

According to a recent study, some 77 percent of the Fortune Global 100 list of companies has at least one Twitter account while the average is 5.8 Twitter accounts, since accounts can easily be set up at the corporate and subsidiary level as well as for specific brands and special events.[7] Thus, Twitter can be an effective strategic marketing tool for Big Oil brands.

The costs of setting up a Twitter account and using it to promote the company's brand and its products are quite low. As with most new media tools, there is a learning curve that companies have to overcome if they want to use them successfully. The impact of using Twitter for marketing purposes can be reduced unless companies make the effort to increase digital word-of-mouth by using the function called "retweeting." Retweeting is the practice of forwarding another person's or company's messages to one's own network of followers. In simple math, the tweet reaches more potential customers. Second, a retweet comes with an endorsement. That is, a third-party is endorsing the brand so the message becomes more honest and transparent rather than an implicit message to buy.

The authors of this study manually collected 1,150 tweets that were a random sampling of tweets between the dates of January 26, 2011 through April 18, 2011. The samples themselves consisted of 47 companies spanning the spectrum of industry, from retail (Whole Foods, Best Buy, and CVS), restaurant chains (such as McDonald's, Starbucks, and Pizza Hut), vehicle manufacturers including (Audi, Chevrolet, and Harley-Davidson), electronics companies (Sony, Nokia, and Samsung), consumer products (Pepsi, Coca-Cola, and Kraft) and entertainment (such as ESPN, MTV, and Disney). Oddly enough, no oil company brands were selected.

After the sampling stage was completed and analyzed, the authors developed a list of 9 Twitter best practices that are paraphrased below:

Don't use all 140 characters at one time. The authors found that Tweets as low as 70 characters long were retweeted nearly twice as often as longer tweets.

Grab the reader's attention. Time-focused messages, like those of the Home Shopping Network, still resonate with consumers. Messages such as "One day only" or "Claim your discount within the next 30 minutes" remind consumers they must act fast if they are to take advantage of a company's offering.

Ask to be retweeted. According to the authors, simply asking for a retweet increased retweeting by 34 percent on average.

Humanize the brand. One of the proven ways to signal to consumers that the brand is more than just selling products and services is to make it a living being. The message can be humorous, a historical view of the brand, or even inspirational.

Announce brand successes. Sharing an accomplishment or award helps to build consumer relationships with the brand. For example, each year *Motor Trend* magazine announces their Car of the Year award as evaluated by their automotive testers and reviewers.

Make tweets educational. In many cases, brands have to educate and inform. Rather than merely push brands on consumers, information that is valued is 51 percent more likely to be retweeted. For example, Exxon Mobil and BP now routinely use Twitter to announce the release of major reports such as their outlooks for energy.

Offer followers a real deal. Messages containing attractive offers are very likely to be retweeted.

Make tweets relevant and current. Topical content tweets are retweeted 41 percent more often than tweets that do not have relevant content.

Act now. Tweets that create a sense of urgency are 24 percent more likely to be retweeted.

Brand lesson no. 14

Big Oil brands were initially slow to assess the effectiveness of social media to build consumer engagement. BP was the first oil company brand to use social media in a crisis communications situation and the success rate was below average. Social media should be used in both crisis communications situations and during normal business operations so that oil company brands engage with consumers.

Notes

1. Sharon Gaudin. "BP Fails to Exploit Web 2.0's Potential," *Computerworld* (June 21, 2010), p. 44.
2. Denise Lenci and John Mullane. "Communicating with the Public: How BP Told the Macondo Story," *Oil & Gas Journal* (December 6, 2010), p. 26.
3. Laura Richardson Walton, Skye C. Cooley, and John H. Nicholson. "A Great Day for Oiled Pelicans: BP, Twitter, and the Deep Water Horizon Crisis Response," *2012 International Public Relations Research Conference.*
4. Ibid.
5. Ibid.
6. Arvind Malhotra, Claudia Kubowicz Malhotra, and Alan See. "How to Create Brand Engagement on Facebook," *Sloan Management Review* (Winter 2013), pp. 18–20.
7. Arvind Malhotra, Claudia Kubowicz Malhotra, and Alan See. "How to Get Your Messages Retweeted." *Sloan Management Review* (Winter 2013), pp. 61–66.

Part V
Concluding Remarks

Abstract: *In 1870, when John D. Rockefeller, Sr. changed the name of the company he founded to Standard Oil, he had not only envisioned and put in place the structure of today's oil industry, but also set in motion something unintended: the negative perception of that industry by the general public and other stakeholder groups. Throughout its corporate existence, Standard Oil used both legitimate and illegitimate methods to win in the marketplace. The illegitimate methods were the most damaging to the Standard Oil brand. Big Oil may see its current use of aggressive marketplace tactics as remaining true to Rockefeller's vision of competition but in today's business environment, the oil industry's poor corporate image remains intact.*

Robinson, Mark L. *Marketing Big Oil: Brand Lessons from the World's Largest Companies.* New York: Palgrave Macmillan, 2014. DOI: 10.1057/9781137388070.0026.

Concluding Remarks

In 1870, when John D. Rockefeller, Sr. changed the name of the company he founded to Standard Oil, he had not only envisioned and put in place the structure of today's oil industry, but also set in motion something unintended: the negative perception of that industry by the general public and other stakeholder groups, including competing firms, local politicians, and state governments. Throughout its corporate existence, Standard Oil used both legitimate and illegitimate methods to win in the marketplace. The illegitimate methods were the most damaging to the Standard Oil brand. Big Oil may see its current use of aggressive marketplace tactics as remaining true to Rockefeller's vision of competition but in today's business environment, the oil industry's poor corporate image remains intact. There are signs, however, that the oil industry's image is changing.

Big Oil's recent marketing campaigns continue to rely on educational themes. Exxon Mobil, using its new "Energy Lives Here" messaging, asks consumers to take their energy quiz. One question asks consumers to identify where most of the United States' oil comes from. The correct answer of course, is the United States, reflecting the country's new sources of Shell oil and gas. Shell Oil's U.S. marketing "Let's Go" campaign continues the themes of helping the environment and engaging with consumers wherever it operates. The messaging is the same; only the medium has changed from traditional media to social media.

Beginning with Rockefeller's adoption of the Standard Oil name in 1870, Big Oil has been around for over one hundred and forty years and its negative image has remained virtually intact. Although it might just take another one hundred and forty years to erase it, I'm waiting to find out.

DOI: 10.1057/9781137388070.0026

Index

Advertorial (advertising-editorial) 89
Adweek 123
Alaska (State of) 7, 45, 49–51, 52, 64, 67, 102, 104
Albany, New York 14
Allen, Thad 138
Alton Railroad 35
Alyeska Pipeline Service Company 52
Amazon Crude documentary 74, 75
American Transfer Company 21
Amoco 26, 30, 65, 69, 95, 97, 98, 100, 101, 103
Andrews Clark & Company 12
Andrews, Samuel 11, 18
Anglo-Persian Oil Company 96
Apple 8
Arab Oil Embargo (1973–1974) 7, 36, 107–110
Archbold, John 21
Atlantic and Great Western Railroad 12, 14
Atlantic Monthly 32
ARCO (Atlantic Richfield Company) 30, 65, 69, 95, 98, 100, 101, 114
Argentina 76
Audi 143

Barron's 90

Baxter, John 68
Beard, Mike 62
Beaumont, Texas 28
Benda, Ernst (Chief Justice of the German Constitutional Court) 60
Benoit, William 33, 48, 50
Benz, Karl 29
Best Buy 143
Best Practice Environmental Option (BPEO) 56, 57, 59, 61
"*Beyond Petroleum*" branding campaign 46, 65, 95, 98, 100–102, 105
Big Oil 2, 3, 6–9, 11, 16, 30, 35–37, 77, 89, 94, 114, 124, 129, 131, 134, 135–137, 142
Black Thursday 18
Bligh Island, Alaska 49
Body Shop 121
Boynton, Judy 82, 85
Bozman, Chris 124
BP 1–3, 6–8, 30, 32, 37, 41, 42–46, 49, 50, 64–70, 95–105, 114, 119, 134, 137–141
BP Conservation Program 102
BP on the Street campaign 95, 98, 99
BP Products North America 65
BrandZ Top 100 Brands ranking 8

Index 149

Brent Spar 55–62, 120, 137
British Department of Energy 57
British Intelligence 97
British Petroleum (see BP)
British Telecom 57
Brown, D.M. (Judge) 76
Brown, John (Lord) 66, 101, 102
Buffalo, New York 13
Bureau of Corporations 34
Burmah Castrol Plc 100, 101, 116
Burmah Oil 95
Burson-Marsteller 130
Business Environmental Leadership Council 102

California 28, 73, 112–114
Caltex 115
Canada 76
Canadian Court of Appeals 76
Canadian oil sands 72
Carnegie, Andrew 96
CBS News
  "*60 Minutes*" news program 74, 75
Central Intelligence Agency (CIA) 97
Chevrolet 143
Chevron 1, 2, 4, 6–8, 26, 30–32, 36, 71–77, 89, 100, 107, 110–117
Chevron at Work 109
Chicago, Illinois 35, 99
Chicago Railroad 35
China National Offshore Oil Corporation (CNOOC) 97
Churchill, Winston 96
Clarian County, PA 21
Clark, Maurice 11
Clean Air Act (1990) 91
Cleveland, Ohio 2, 10–15, 18, 19, 23, 32
Cleveland Dealer newspaper 11
Cleveland Leader newspaper 13
Cleveland Massacre 18
Climate, Community, and Biodiversity Alliance 102
Coast Guard (U.S.) 43, 49, 50, 138, 140
Coca-Cola 143
Columbia Journalism Review 75
ConocoPhillips 30, 100, 101

Continental Oil (see ConocoPhillips)
Costco 93, 131
Count on Shell advertising campaign 121, 122
CNN 124
CVS 143

Daimler, Gotlieb 29
D'Arcy, William Knox 96
Darley, John 81
Department of Justice (U.S.) 43
Deepwater Horizon rig 1, 3, 41–43, 49, 50, 64, 104
D'Esposito, Steve 58
Discovery cable channel 124
Disney 143
Donziger, Steven 76
Drake, Col. Edwin 11
Dudley, Robert (Bob) 45, 102
Duryea, Frank and Charles 29

E! cable channel 124
East Saint Louis, Illinois 35
Economist Group 116
Ecuador 36, 71–77, 97, 116
Eggart, Tim 57
Elkins Act (1903) 35
Empire Transportation Company 20
Energy and Biodiversity Initiative 102
Environmental Defense 102, 104
Environmental Protection Agency (U.S.) 140
ESPN 124, 143
Erie Railroad 14
Esso (Eastern Seaboard Standard Oil) 29
Esso AG 56
European Commission 65, 69
Excelsior Works refinery 12
Exxon 23, 26, 29–31, 34, 36, 42, 45, 48–53, 65,73, 89, 100, 114, 119, 137, 138, 141
Exxon Mobil Corporation 1, 2, 6–8, 23, 24, 29, 32, 56, 100, 115, 144, 147

DOI: 10.1057/9781137388070.0027

150   Index

Exxon *Valdez* oil tanker   1, 7, 42, 45, 48, 49, 52, 65, 137

Facebook   4, 46, 136, 137, 139, 141–143
Fay, Christopher   61
Financial Times of London newspaper   8, 32
Fitch (corporate ratings service)   69
Flagler, Henry M.   13, 16, 18, 21
Flickr   139
Fortune magazine company rankings   7, 100, 143
Friendly Serve (Mobil Oil)   4, 93, 131
Ford, Henry   29
Ford Motor Company   119
Fourth International North Sea Conference   59
Fuelmatics   134

Gallup (polling service)   7
General Petroleum   29
Generation F3 drivers   92
Geological Survey (U.S.)   43
Gilligan, Dan   124
Golf Digest magazine   124
Google   8
Greenpeace   55, 58–62, 101, 103, 104, 127
Greenwash Award   101
Gulf of Mexico oil spill   7, 42, 45, 46, 50, 64, 102, 104, 116, 137, 140
Gulf Oil   111

Hamilton, Martha M.   75
Harkness, Stephen   18
Harley-Davidson   143
Harrisburg, Pennsylvania   14
Harrison, Benjamin (U.S. President)   32
Harvick, Kevin   124
Hayward, Tony   42–47, 50, 139
Hazelwood, Joseph   49, 50
Henry, Simon   81
Hewitt & Tuttle   11
History of the Standard Oil Company   2, 27
Hofmeister, John   9, 125

Home Shopping Network   144
Homebodies drivers   92
House Energy and Commerce Committee (U.S.)   37, 67
Human Energy advertising campaign   4, 107, 108, 116, 117
Hurricane Katrina   141

Institute of Public Relations   62
Interstate Commerce Committee   28
International Climate Change Partnership   102
International Maritime Organization (IMO)   56
Iran   7, 96, 97, 107, 108, 110

J. Walter Thompson (advertising agency)   112, 113, 121, 123, 124
Jacobi, Mary Jo   81
Jennings, O.B.   18
Joint Nature Conservancy Committee   57
Johnson & Johnson   46
  Tylenol brand crisis   46

Kaplan, Lewis A. (Judge)   75–77
Keker & Van Nest LLP   76
Kenney, John   105
Kerley, Richard   137
Kerosene   10, 12, 13, 16, 22, 23, 26
Kilgore, Joe   122
Kohl, Helmut   60
Kraft   143
Kuwait Petroleum   97
Kyoto Protocol   102

Lake Erie, PA   14
Lake Shore Railroad   14, 21
Landis, Kenesaw Mountain (U.S. judge)   35
Landor & Associates   35, 36
Landry, Mary   138
Latin America   72
  Ecuador   36, 71–77, 116
  Venezuela   72
Leech, Kathy   98, 99

DOI: 10.1057/9781137388070.0027

Let's Broaden the World's Energy Mix (Shell advertising tagline) 127
Let's Go (Shell advertising campaign) 126, 127, 147
Levick Strategic Communications 137
London (U.K.) 13, 99
Lorfelder, Jochen 39, 59
Lynch, Gerard E. 75

Macondo oil well 42, 138
Magnolia Petroleum 29
Major, John (U.K. Prime Minister) 60
Manzoni, John 66
Maryland (U.S. state) 22
McClure's magazine 2, 27, 32, 33
McCollum-Spielman (market research firm) 109–113
McDonald's 119, 143
McGarryBowen
McKinley, William (U.S. president) Assassination 27
Melchett, Lord Peter 61
Merritt, Carolyn 65
Microsoft 8
Mobil Oil Corporation 3, 23, 29, 42, 88, 89, 92, 93, 135, 138
 advertorials 3
 speedpass 3
Moody-Stuart, Mark 124
Morgan, J.P.
Mossadeq, Mohammad 96
MTV 149

National Association of Stock Car Auto Racing (NASCAR) 115, 124
National Court of Justice (Ecuador) 77
NASCAR Scene 124
National Dialogue on Energy Security (Shell Oil U.S.) 125
National oil companies 80, 114
Nature magazine 61
New Jersey (U.S. state) 22, 24, 27
New York Central Railroad 14, 20, 21
New York City 2, 3, 13, 14, 23
New York Stock Exchange 104

New York Times newspaper 33, 89–92, 105
Nigeria 72, 120, 121
Nokia 143
Nordstrom 133
North Sea 56, 58, 61, 120
NPD (market research firm) 115

Obama, Barrack (U.S. president) 44, 45
Ogilvy & Mather (advertising agency) 101, 121
Ohio Supreme Court 24, 27
Oil City Derrick newspaper 19
oil discoveries 27
 California 27
 Illinois 27
 Oklahoma 27
 Spindletop, TX 27
oil prices 12, 14
Oil Regions (Pennsylvania) 14

Papua Conservation Fund 102
Paris, France 13
Penn, Schoen & Berland 126
Pennsylvania (U.S. state) 11, 14, 18, 19, 21, 22
Pennsylvania Railroad 14, 20
People Do advertising tagline 112–114, 116, 117
Pep Boys (retail store) 94
Pepsi 143
Petrobras 97
PetroEcuador 72, 73, 75, 97
Pew Trust 102
Petroleum Act (U.K.) 57
Petroleum Marketers Association 124
Philadelphia, Pennsylvania 14, 18
Pittsburg, Pennsylvania 14, 18, 113
Phillips Petroleum 100
Pizza Hut 143
Popular Mechanics magazine 124
Pratt, C.M. 34
Prince William Sound (Alaska) 49, 52
price driven consumers 92

Index

Profits and Principles: Is There a Choice? (Royal Dutch Shell ethics report) 120, 121
Prudhoe Bay, Alaska (oil field) 67, 68, 104
Public Opinion Monitor (Internal Chevron report) 111

Ranseur, David 51
Rawl, Lawrence 48, 51–53
Rexrodt, Gunter (German Economics Minister) 60
Richford, New York 11
Roddick, Anita 121
Road Warrior drivers 92, 93
Rockefeller, Andrews and Clark 12
Rockefeller, Andrews and Flagler 13, 18
Rockefeller and Company 13
Rockefeller, Sr, John D. 1–3, 10–16, 18–24, 27–29, 31–38, 73, 77, 96, 146, 147
Rockefeller, William 18, 23
Roden, Danny
Roosevelt, Theodore (U.S. president) 27, 34, 36
Royal Dutch Petroleum Company 118, 119
Royal Dutch Shell 1–3, 6–8, 30, 32, 56, 57, 69, 79, 80, 118, 119
Royal Dutch Shell (Germany) 57, 59, 60
Royal Dutch Shell (U.K.) 56, 57, 59–61
Russia 72

Samsung 143
Sangeorge, Robert 61
Saudi Aramco 97
Schmertz, Herbert 91
Schultz, Herbert 132
Scottish National Heritage 57
Senate Permanent Subcommittee on Investigations (U.S.) 36
Sheetz 93, 94, 131
Shell Oil (U.S. subsidiary of Royal Dutch Shell) 2, 7, 9, 30, 32, 93, 118, 120–122, 124, 125, 147

Answer Man campaign 120
Let's Go campaign 4, 126, 127, 147
Sherman Antitrust Act (1890) 28, 32
Shell Transport and Trading 30, 81, 82, 85, 118, 119
Sierra Club 104
Socony Mobil 29
Socony Vacuum 29
Solarex 101
Sony 143
South Improvement Company 19
Speedpass (electronic device; Mobil Oil) 4, 93, 131
Spindletop, Texas (oil discovery) 28
Springfield, Massachusetts 29
Sports Illustrated magazine 124
Standard Oil Company (Ohio) 2, 9, 11, 16, 18, 22, 27, 31, 89, 146, 147
Standard Oil Company of California 30, 73, 89, 107, 108
Standard Oil of Indiana (Amoco) 30, 35, 36
Standard Oil of New Jersey (Exxon) 23, 24, 27–29
Starbucks 132, 143
Statoil 69
Stone, Amasa 21
Story of a Great Monopoly (Atlantic Monthly article)32
Standard Oil of New York (Mobil Oil) 22, 29
Standard Oil Trust 1, 22–24, 26, 27, 32, 38
Standard Oil Works refinery 12
Stevens, William 53
Stone, Amasa 121
Strongsville, Ohio 11
Superior Court of Justice (Canada) 76
Svanberg, Carl-Henric 44, 45, 47

Tarbell, Ida 2, 3, 27, 32, 33
Texaco 30, 71–75, 100, 114–117
Texaco Petroleum (Texpet) 71–73
Texas City, Texas refinery 41, 64–66, 102, 104
Trans-Alaska pipeline 68, 104

True Blues drivers 92, 93
Trusts: The Recent Combination in Trade 24
Twitter 4, 46, 136–137, 139–144

United Nations Commission on International Trade Law 74
United Pipelines Company 21
Untapped Energy (Chevron television commercial) 116
U.S. Chemicals Safety and Hazard Investigation Board (CSB) 66
U.S. Commodity Futures Trade Commission (CFTC) 67
U.S. Court of Appeals 76
U.S. District Court (Manhattan) 74
U.S. District Court (St. Louis) 28
U.S. Environmental Protection Agency (EPA) 140
U.S. Occupational Safety and Health Administration (OSHA) 66
U.S. Second Court of Appeals 76
U.S. Securities and Exchange Commission (SEC) 80, 83
U.S. Supreme Court 26, 29, 76

Vacuum Oil Company 29
Valdez, Alaska 49
Valdez Harbor, Alaska 52
Van de Vijver, Walter 81, 85

Vanderbilt, Cornelius 96
Vanderbilt, Jacob J. 21
Vanderbilt, William H. 21
Vanderbilt and Forman 21
Venezuela 72

Wade, Mark 121
Wall Street Journal 32, 90, 139
Wal-Mart 133
Walter Thompson, J. (advertising agency) 112, 113, 121, 123, 124
Warner, Rawleigh 91
Washington, DC 53, 99
Watson, David K. 24
Wawa 93, 94, 131
Watts, Sir Philip 81, 82, 85
We Agree Chevron advertising tag line 107, 108, 116, 117
Whiting, Indiana 35
Wildman, Karen 123
Woollam, Richard 37, 68, 69
World Resources Institute 102
Worldwide Fund for Nature 61

Yes Men 127
YouGov Brandweek Buzz Report 42
YouTube 4, 46, 127, 136, 137, 139

Zindler, Harald 59, 60